Para Jennie
my colleague
mating Earth foods
from GR

# W I L D
# N e s s

Thanks so much!
Michael

For DeMarco,
welcome to Gunnison's Wildness!
In gratitude,
Robin Wall Kimmerer

For
Jennie
with love
Brooke Hunt

to wildness,
Sam —

Aaron Abeyta

# W I L D
# N e s s

*Relations of People & Place*

*Edited by Gavin Van Horn & John Hausdoerffer*

THE UNIVERSITY OF CHICAGO PRESS * CHICAGO AND LONDON

PUBLISHED IN COLLABORATION WITH THE CENTER FOR HUMANS
AND NATURE

The University of Chicago Press, Chicago 60637
The University of Chicago Press, Ltd., London
© 2017 by The University of Chicago
All rights reserved. No part of this book may be used or reproduced in any manner whatsoever without written permission, except in the case of brief quotations in critical articles and reviews. For more information, contact the University of Chicago Press, 1427 E. 60th St., Chicago, IL 60637.
Printed in the United States of America

26 25 24 23 22 21 20 19 18 17    1 2 3 4 5

ISBN-13: 978-0-226-44466-6 (cloth)
ISBN-13: 978-0-226-44483-3 (paper)
ISBN-13: 978-0-226-44497-0 (e-book)
DOI: 10.7208/chicago/9780226444970.001.0001

Library of Congress Cataloging-in-Publication Data

Names: Van Horn, Gavin, editor. | Hausdoerffer, John, editor.
Title: Wildness : relations of people and place / edited by Gavin Van Horn and John Hausdoerffer.
Description: Chicago : The University of Chicago Press, 2017. | Includes bibliographical references and index.
Identifiers: LCCN 2016033175 | ISBN 9780226444666 (cloth : alk. paper) | ISBN 9780226444833 (pbk. : alk. paper) | ISBN 9780226444970 (e-book)
Subjects: LCSH: Wilderness areas. | Wilderness areas—United States. | Nature—Effect of human beings on. | Wildlife conservation.
Classification: LCC QH75 .W5273 2017 | DDC 333.78/2—dc23 LC record available at https://lccn.loc.gov/2016033175

♾ This paper meets the requirements of ANSI/NISO Z39.48-1992 (Permanence of Paper).

# Contents

*Acknowledgments ix*

Introduction: Into the Wildness
*Gavin Van Horn*     1

PART 1. WISDOM OF THE WILD     9

1. Wildfire News
   *Gary Snyder*     11

2. Conundrum and Continuum: One Man's Wilderness, from a Ditch to the Dark Divide
   *Robert Michael Pyle*     12

3. No Word
   *Enrique Salmón*     24

4. The Edge of Anomaly
   *Curt Meine*     33

5. Order versus Wildness
   *Joel Salatin*     43

6. Biomimicry: Business from the Wild
   *Margo Farnsworth*     50

7. Notes on "Up at the Basin"
   *David J. Rothman*                                60

PART 2. WORKING WILD   65

8. Listening to the Forest
   *Jeff Grignon and Robin Wall Kimmerer*           67

9. The Working Wilderness
   *Courtney White*                                 75

10. The Hummingbird and the Redcap
    *Devon G. Peña*                                 89

11. Losing Wildness for the Sake of Wilderness:
    The Removal of Drakes Bay Oyster Company
    *Laura Alice Watt*                              100

12. Inhabiting the Alaskan Wild
    *Margot Higgins*                                113

13. Wilderness in Four Parts, or Why We Cannot
    Mention My Great-Grandfather's Name             123
    *Aaron Abeyta*

PART 3. URBAN WILD   135

14. Wild Black Margins
    *Mistinguette Smith*                            137

15. Healing the Urban Wild
    *Gavin Van Horn*                                145

16. Building the Civilized Wild
    *Seth Magle*                                    156

17. Cultivating the Wild on Chicago's South Side:
    Stories of People and Nature at Eden Place
    Nature Center
    *Michael Bryson and Michael Howard*                166

18. Toward an Urban Practice of the Wild
    *John Tallmadge*                                  177


   PART 4. PLANETARY WILD    *187*

19. The Whiskered God of Filth
    *Rob Dunn*                                        189

20. The *Akiing* Ethic: Seeking Ancestral Wildness beyond
    Aldo Leopold's Wilderness
    *John Hausdoerffer*                               195

21. On the Wild Edge in Iceland
    *Brooke Hecht*                                    205

22. The Story Isn't Over
    *Julianne Lutz Warren*                            214

23. Cultivating the Wild
    *Vandana Shiva*                                   228

24. Earth Island: Prelude to a Eutopian History
    *Wes Jackson*                                     233

    Epilogue
    Wild Partnership: A Conversation
    with Roderick Frazier Nash
    *John Hausdoerffer*                               243


                *Permissions* 255
              *About the Contributors* 257
                   *Index* 265

# Acknowledgments

The Mi Casita cabin in Tres Piedras, New Mexico, is a place of powerful intersections. The cabin sits at the ecological intersection of the subalpine forests of the San Juan Mountains and the high semideserts above the Rio Grande, creating a rich edge habitat. It also lies at the cultural intersection of multigenerational Hispano families and Taos Pueblo communities and more recent Anglo ranchers, international tourists, and federal land managers. In short, the site is a complex blend of relations between land and people.

But there is one more intersection that is highly relevant to this book, *Wildness*. In 1912, conservationist Aldo Leopold, newlywed and newly appointed supervisor of the Carson National Forest, moved into Mi Casita with his wife, Estella. Here, Leopold's thinking shifted from that of a Pinchot-trained utilitarian forester to a holistic land manager as he began asking his Forest Service colleagues to consider the "whole loaf" of the forest—soil, grass, wildlife, recreation, water, as well as timber. This holistic shift, for Leopold and the modern environmental movement that followed, eventually led him to advocate for New Mexico's Gila Wilderness, which became the first designated wilderness area in 1924. But it also led him to declare that "the weeds in a city lot convey the same lessons as the redwoods"—that the health of land, the wildness embedded in land's ability to provide life for all beings and energy for all systems, rests in its "capacity for self-renewal."

In the summer of 2015, we (Gavin and John) visited Mi Casita to edit and organize the final essays of this volume. Sitting on the porch of Mi Casita, looking east to the high mountain edge of the Rio Grande watershed, we took in the view of afternoon lightning storms gathering above the Sangre de Cristo Mountains and could not help but reflect upon how—a century after Leopold lived in this cabin—this place continues to inspire intersections. Mi Casita provided a perfect setting for finishing *Wildness*, a merging together of so many individual voices, organizational benefactors, and community partners. The two of us would like to thank those who made this book possible and to recognize the intersections that led to (and we hope will lead out of) this book.

Gavin Van Horn bows first in gratitude to his small wild clan, Marcy and Hawkins, who provide him the laughter and love that sustains him in everything else. They graciously allowed him to spend three weeks at Mi Casita as a writer-in-residence for the Aldo and Estella Leopold Residency Program, which contributed a great deal toward the final shape of this project. Gavin would like to recognize the sponsors of the Mi Casita program—the Aldo Leopold Foundation and the US Forest Service—as well as Chris Furr and Tony Anella in particular, who made his stay so enriching. The smell of piñon lingers in his memory. Gavin also gives thanks to his amazing colleagues—James Ballowe, Hannah Burnett, Anja Claus, Kate Cummings, Emilian Geczi, Brooke Hecht, Bruce Jennings, Curt Meine, and Jeremy Ohmes—truly astounding people who keep his mind sharp and his heart full.

John Hausdoerffer offers his deepest thanks to his partner, Karen, and to the energizing brightness of his daughters, Atalaya and Sol. John thanks his colleagues at Western State Colorado University for supporting the sabbatical that allowed him to research his chapters and visit thinkers and authors who are featured in *Wildness*, such as Winona LaDuke, Vandana Shiva, and Devon Peña. John would like to express his gratitude to those hosts and their respective organizations for their hospitality: the White Earth Land Recovery Project, the Navdanya Institute, and the Acequia Institute. John is also humbled by the opportunity to coedit this book as a Fellow for the Center for Humans and Nature, an institution proving that public discourse can still be driven by compelling questions of enduring relevance. This book would not have been possible without Western State's Headwaters Conference, which allowed John to host and connect with *Wildness* authors Gary Snyder, Aaron Abeyta, Enrique Salmón, Laura Watt, Michael Bryson, Gavin Van Horn, and John Tallmadge. Special thanks go to the Aldo Leopold Foundation for making John a Visiting Fellow in 2013 and to the Aldo and Estella Leopold Residency Program coordinated by the US Forest Service and the Aldo Leopold Foundation for his time as a writer-in-residence in 2013.

Together, John and Gavin would like to thank the board and staff of the Center for Humans and Nature, who made this book project possible. The excellence of the initial brainstorming meeting in Crested Butte, Colorado, was due to the center's logistical and financial support, and the project and the conversations benefited tremendously from a dose of mountain air. Deep thanks go to editorial director Christie Henry and the University of Chicago Press. Christie has been unflagging in her encouragement and belief in the book; to work with her is a delight. The team at the press, including designer Lauren Smith and editorial associates Gina Wadas and Caterina MacLean, has been fantastic. Copyeditor Kyriaki Tsaganis of Scribe Inc. as well as indexer Jim Fuhr also deserve recognition for their truly superb work and attention to detail. Lastly, to all the people who contributed their stories to this book, it was humbling to work with such expert wordsmiths and, more important, such wonderful human beings. It is your unique perspectives about wildness and the work of communities that nurture wildness that give us hope.

Wildly,
Gavin and John

\* INTRODUCTION \*

# INTO THE WILDNESS

*Gavin Van Horn*

I am standing a few feet off the edge of a manmade canal. A great blue heron is frozen in contemplation of the shallow waters around his ankles. Cottonwood and ash trees display the toothy signs of beaver. Further up the canal, a pair of coyotes readies for dusk. I graze the tips of my fingers across the ivory bloom of an unassuming flower, thinking of the persistence of the land's memory and how it may feed our future.

In Robert Macfarlane's *The Wild Places*, he highlights a Chinese artistic tradition known as *shan-shui* ("rivers and mountains") that began in the fifth century BCE. He writes, "Their art, like that of the early Christian monks, sought to articulate the wondrous processes of the world, its continuous coming-into-being. To this quality of aliveness, the *shan-shui* artists gave the term *zi-ran*, which might be translated as 'self-ablazeness,' 'self-thusness' or 'wildness'" (31).

Wondrous processes of the world. Coming-into-being. Self-ablazeness. Thusness. Wildness. Great blue heron. Cottonwood. Ash. Beaver. Coyote. Human. Together, in and along a city canal, with an unassuming flower. In 1942, wildlife ecologist Aldo Leopold delivered a pithy presentation at the seventh North American Wildlife Conference, exploring how ecology professors could engage all their students in seeing, understanding, and enjoying the land. His goal, as was the case in many of his writings, was to impart the "concept of land-health[,] the sustained self-renewal of the community," an exploration he thought should be pursued well beyond formal educational settings. This self-renewing community, for Leopold, included humans. He put this plainly: "Who is the land? We are, but no less the meanest flower that blows. Land ecology discards at the outset the fallacious notion that the wild community is one thing, the human community another" (303).

The book in your hands takes this idea as a starting point: that human and wild communities are entangled and can work toward collective health and self-renewal—toward self-ablazeness.

## *Circling Wildness*

There is no singular meaning for wildness. The most prevalent definitions of *wild* usually refer to the etymological roots of the term, which are associated with "self-will." Whether it is a place, a nonhuman animal or plant, or a state of mind, *wild* indicates autonomy and agency, a will to be, a unique expression of life.

Those of us who identify as conservationists or as people concerned with the well-being of the natural world are familiar with the positive connotations for wild. We've inherited Leopold's call for a shift of values toward "things natural, wild, and free"; we nod our heads when we hear Thoreau's famous dictum, "in Wildness is the preservation of the world"; we may link wildness to collective political ideals, as Terry Tempest Williams does when she describes the "open space of democracy"; and we likely read approvingly of what the wild can impart to us, from persons such as Buddhist poet and essayist Gary Snyder:

> We can appreciate the elegance of the forces that shape life and the world, that have shaped every line of our bodies—teeth and nails, nipples and eyebrows. We also see that we must try to live without causing unnecessary harm, not just to fellow humans but to all beings. We must try not to be stingy, or to exploit others. There will be enough pain in the world as it is.
>
> Such are the lessons of the wild.... Wildness is not just the "preservation of the world," it *is* the world. (168–69)

As many of the contributors to this volume point out, however, we should be aware that culture informs our usage of words and the meanings we ascribe them. *Wild* is no exception. One common set of connotations for *wildness* is that which is forbidden, dangerous, or out of control. When the rock band Steppenwolf belted out "Born to Be Wild," they offered up an anthem to rebellion, independence, and middle-fingered freedom. "Like a true nature's child," we are born to be wild—to escape social confinements and conventions, to go feral, apparently while riding a motorcycle into the sunset.

The point is that *wild* is not always thought of as a virtue, or virtuous for that matter. The meaning of "self-will" often bears with it a particular set of assumptions about the autonomous *self*. This focus on the wild *self* sometimes frames wildness in terms of individual pursuit and actualization; it implies that nature is a means to an individual end, something to exploit or ride across on your way to freedom.

To perceive wildness in this way is not right or wrong, but it does reflect a truncated view of what constitutes *self-will*. Take, for example, an individual acorn. True, a single acorn, given the right set of conditions, becomes a single oak tree. But the oak tree needs soil, air, water, sunlight, animal seed dispersers and pollinators, possibly fire, and so on. In other words, the wildness of a single oak tree is actualized through the wild processes of a community. Absolute freedom from community is to be isolated; carried to its conclusion, such wildness is pathological and ultimately *self-destructive*.

Another destructive meaning for *wild*, as several of our contributors point out, is that the word has been wielded as a blunt instrument, a way to characterize peoples and places as *things* apart, a linguistic wedge that separates out that which is deemed less than human, usually as justification for conquest. As Enrique Salmón points out in chapter 3, for native peoples such as the Rarámuri, "in [their] persons, often described as savage or beastly, the European gaze saw only extensions of their [own] concept of a new wilderness," which made the term "a sociolinguistic construction that had real consequences." Wild places and peoples, in this conceptual framework, are the Other—to be feared and, if possible, subdued.

Returning to Gary Snyder's words for the moment, we might notice that when he speaks of wildness, he does not speak exclusively of wild places. Snyder speaks more comprehensively of wild *processes*, "the forces that shape life and the world, that have shaped every line of our bodies." He distinguishes between wilderness-as-place, set apart, and wildness-as-practice, the process of becoming respectful coinhabitants of place. A practice of the wild implies that humans have a role to play in cultivating wildness—in themselves, in their communities, and in the landscapes of which these communities are a part. Wildness, in this sense, is an ongoing relationship, one in which human cultures—through active participation and humble restraint—become attuned to the community of life and constitutive of the well-being of the places in which they live, work, and play.

## The Relative Wild

Two contrasting ideas about nature are widely held in contemporary contexts. One is that nature is most wild in the absence of human presence, that wildness can only truly exist *apart* from the defiling hands of humans. The other is that nature is completely humanized, that nothing is untouched by human activity, from the oceans below to the sky above. One conclusion that can be drawn from a no-analog future is that further manipulations, whether genetic, ecological, or climatic, are not only necessary but justified. The political and social implications of these divergent perspectives can create heated discussion, and fresh fuel got tossed into the campfire during the run-up to the fiftieth anniversary celebration of the Wilderness Act (see, in particular, Minteer and Pyne 2015; Wuerthner et al. 2014; Marris 2011; Nelson and Callicott 2008). A good deal of the discourse—and the disagreements—involves the magnitude of change we currently see in our world and if human participation in that change is deemed heroic, benign, or pernicious.

Important as it may be to discuss and debate these assessments of nature and human nature, our approach to this book takes a different path, sharing stories *across the wild continuum*. The essays celebrate *degrees* of wildness that exist on this continuum, and the possible ways in which human communities can nurture, adapt to, and thrive with wildness. Wildness is not an all-or-nothing proposition. There are variations, ranging from the sunflower pushing though a crack in a city alley to the cultivated soils of a watershed cooperative to thousands of acres of multigenerational forestlands. Wildness can be discovered in everyday places, from one's backyard to one's bioregion, from holistically managed rural ranches to reclaimed urban industrial sites, from microscopic to atmospheric scales. Each author holds up a vision of wildness through stories about particular people in particular places.

As much as physical locations animate these stories, however, wildness is not simply an external goal—a place we protect, a landscape we honor, or the ecosystem complexity we restore. We also carry wildness within. We are *related*—relatives of—other plants and animals (and fungi and algae and protozoa and so on) who inhabit this earth with us, and some of whom inhabit this earth on and in us. We are kin. This applies generally: as human beings, we share an evolutionary and ecological kinship, a family tree whose branches reach through place and time, knitting us together on a common journey. This also applies particularly: there are certain species and places with whom we share deep and meaningful relationships on personal and cultural levels. The stories of how our lives fit with theirs, including what it means to be fully human in their presence, emerge out of our observations, engagements, and codependencies with these wild nonhuman kin.

These two themes—degrees of wild relatedness across landscapes and depth of wild kinship in place—set the tone for each contributor's exploratory journey into what wildness is, what it could be, and how it might be recovered in our lives.

## Wild Journeys

We've brought together many authors, from a variety of landscapes, cultures, and walks of life, to tell their stories about the interdependence of wildness and human lifeways. As we looked over the larger landscape of the book, the essays seemed to group themselves into four categories: Wisdom of the Wild, Working Wild, Urban Wild, and Planetary Wild.

Part 1, "Wisdom of the Wild," takes the reader on a search for what we can learn from wild species and systems, including how our own lives and livelihoods can be respectfully integrated with our landscapes. Chapter 1 offers a poetic salutation for the book, a nod toward chaotic planetary forces and the embodied wisdom of adaptive organisms such as redwood trees—a wild dance millions of years in the making that places human achievement in perspective. In chapter 2, we find a celebration of activism for "the entire blessed continuum," from the small wild discoveries of a suburban Colorado irrigation ditch to the capital-W *Wild* of the Dark Divide Wilderness Area. "Kincentric landscapes" of the Rarámuri peoples are the focus of chapter 3, in which humans play a keystone role in enhancing central Mexico's biological diversity, but for whom the word *wild* is laced with colonial meanings. In chapter 4, we delve into the historical and contemporary revolution in the Driftless region of Wisconsin, whose inhabitants have learned to "turn" with those contoured lands. Chapter 5 takes us to Polyface Farm in northwestern Virginia to discover how an "ecofarmer" submits to wildness. In chapter 6, we meet business people seeking out the genius of place in west-central Georgia, finding inspiration from wild systems that have become muse, teacher, and template for their company's practices. And in chapter 7, we visit Massachusetts' mountain slopes, where the greater wildness of which we are always already a part shapes the poetic imagination even as it raises difficult questions.

In the "Working Wild" (Part 2), we learn about the relationality of wildness. In chapter 8, we get a bird's-eye as well as an understory view of the forests of northern Wisconsin, where regeneration forestry is alive and well among the Menominee, Maple, and Pine nations. We witness the interactions of ranchers and the land in chapter 9, learning lessons from both sides of the fence of holistically managed ranchlands in the American Southwest. Chapter 10 offers the perspective of the ruby-throated *colibrí* as he revels in "arid-sensible" acequias of southern Colorado. In chapters 11 and 12, we hear the stories of a "potential wilderness" in a central California tideland and an officially designated wilderness area in Alaska, both of which include policy provisions for small-scale human uses—or did until recently—creating interesting tensions between wilderness ideals and relatively wild, sustainable local livelihoods. And in chapter 13, we follow sheep, and the men who are responsible for their lives and deaths, into the Toltec wilderness, where family and prayer, memory and landscape intertwine. All these stories reveal the possibilities of wildness thriving not in spite of, but with—and sometimes even because of—human economies that are attuned to the places in which they are embedded.

Part 3 draws our attention to the "Urban Wild" and the ways in which cities are not domesticated, lifeless, throwaway lands but full of wild creatures, habitats, and possibilities. We learn of black "ecotones" in chapter 14, where joyful relationships to land are not defined by the bonds of ownership but by a strong sense of freedom and rootedness. In chapter 15, we venture to the South Side of Chicago to hear about a dangerous wildness while considering how ecological restoration might cultivate mutual healing for both land and people. We ride through the city with an urban ecologist in chapter 16 into one of "the last great unexplored ecosystems," discovering adaptive animals and how we might adapt better to them. In chapter 17, we witness how reclaimed land, even in some of the most neglected and economically disadvantaged urban areas, can become a new Eden for children and their families to experience the wonder of recreated wildness. Chapter 18 offers possibilities of a *practice* of the wild in places like Cincinnati through the disciplines of mindfulness, attentiveness, husbandry, pilgrimage, and witness.

In the final section of the book, "Planetary Wild" (Part 4), imaginations run free, roaming throughout the world and into wild futures in the making. Chapter 19 introduces us to the colobus monkey-deities of West Africa and the people who honor and protect their sacred groves, as well as the decomposing gods of rebirth who make all filth wild again. Chapter 20 places us in a canoe on a Minnesota lake during the Anishinaabeg rice harvest, seeking answers to the question, "What kind of ancestor do you want to be?" In chapter 21, we travel to elven churches in Iceland, finding surprises at the wild margins of beleaguered forests. In chapter 22, we ponder the meaning of the Anthropocene, our fate and nature's entwined, and the possibilities for a wild reborn out of a climate in chaos. We consider the wild "self-organizing" power of Shakti in chapter 23, and the cocreation of living energies in the struggle to reclaim traditional farming in India. The volume closes with two different visions of the future—one close at hand (chapter 24), one a thousand years from now (epilogue)—in which humans have learned to respect wild systems and creatures. The first "eutopia" considers new paradigms for agriculture, looking to wild systems as a standard for creativity, complexity, and diversity; the second is a conversation about the merits of a future "Island Civilization" and whether humans can be "housemates" or merely remote neighbors when the "self-will" of the land is the ultimate priority.

What the reader will find across these diverse geographies of the wild continuum, from the near-at-hand to the planetary, are experiences grounded in everyday lifeways that respect, learn from, and provide for the agency of nonhuman beings. These stories offer models for living with and becoming more. You will find an understanding of humans as "keystone species"—not just that we change things, as do all creatures, but how we might do so reflectively, responsibly, and responsively. You will hear about wildness in terms of not only place but also deep time. And you will read about wild *process*, a dance that creates possibilities for the dynamic flourishing of life, an evolutionary project to which we are bound and to which we can contribute.

In all the stories in this volume, we find ways in which the flames of wild "self-ablazeness" are being fanned. They beckon us onward, to discoveries and recoveries of wildness.

## *References*

Leopold, Aldo. "The Role of Wildlife in Liberal Education." In *The River of the Mother of God and Other Essays by Aldo Leopold*, edited by J. Baird Callicott and Susan L. Flader, 301–6. Madison: University of Wisconsin Press, 1991 [1942].

Macfarlane, Robert. *The Wild Places*. New York: Penguin Books, 2007.

Marris, Emma. *The Rambunctious Garden: Saving Nature in a Post-Wild World*. New York: Bloomsbury, 2011.

Minteer, Ben A., and Stephen J. Pyne. *After Preservation: Saving American Nature in the Age of Humans*. Chicago: University of Chicago Press, 2015.

Nelson, Michael P., and J. Baird Callicott. *The Wilderness Debate Rages On: Continuing the Great New Wilderness Debate*. Athens: University of Georgia Press, 2008.

Snyder, Gary. *The Practice of the Wild*. New York: North Point Press, 1990.

Thoreau, Henry David. "Walking." In *Henry David Thoreau: Collected Essays and Poems*, edited by Elizabeth Hall Witherell, 225–55. New York: Library of America, 2001.

Williams, Terry Tempest. *The Open Space of Democracy*. Great Barrington, MA: Orion Society, 2004.

Wuerthner, George, Eileen Crist, and Tom Butler. *Keeping the Wild: Against the Domestication of Earth*. Washington, DC: Island Press, 2014.

# PART 1

*Wisdom of the Wild*

# 1

## WILDFIRE NEWS

*Gary Snyder*

For millions,
for hundreds of millions of years
there were fires. Fire after fire.
Fire raging forest or jungle,
giant lizards dashing away
big necks from the sea
looking out at the land in surprise—
fire after fire. Lightning strikes
by the thousands, just like today.
Volcanoes erupting, fire flowing over the land.
Huge Sequoia two foot thick fireproof bark
fire pines, their cones love the heat
how long to say,
that's how they covered the continents
ten lakhs of millennia or more.

I have to slow down my mind.
slow down my mind
Rome was built in a day.

# 2

## CONUNDRUM AND CONTINUUM

*One Man's Wilderness, from a Ditch to the Dark Divide*

*Robert Michael Pyle*

Conundrum Hot Springs bubbles out of a slice of alpine fairyland in Colorado's Maroon Bells-Snowmass Wilderness Area. This modest-sized hot spring, nested in the lap of high meadows thick with magenta paintbrush and Colorado blue columbines, is situated at 11,200 feet above sea level, a good, stiff hike from any trailhead. When I got to know it in the mid-1970s, hiking from the Rocky Mountain Biological Laboratory at Gothic to Aspen, it was in good shape and never very crowded with the naked soakers who came to supplicate its powers. That changed with popularity, as more weekend sybarites made the long trek up from Aspen below. Internet images now show Conundrum still pretty, not entirely spoiled, but sometimes bank-to-bank with bodies, as if it were the Blue Lagoon in Iceland. The US Forest Service has had to impose various controls on numbers and camping in order to protect the sensitive site. Conundrum Hot Springs is, relatively speaking, still wild. But it has grown less Wild.

What do I mean by that? The words "wild" and "wilderness" have different connotations for different people. They can stand for anything you love, anything you don't understand but are trying to, or anything you long for, fear, resent, or rejoice in; they stand for any state you wish to elevate or derogate, praise or blame, inveigh against or agitate for. It would seem that these words can mean almost anything you want them to. Are they therefore useless as words? As ideals or as realities? I don't think so. But I believe we should take great pains to be clear when we discuss them, and even more so when we seek to apply them on the land.

The chief conundrum here, as I see it in the context of our ravenous contemporary culture, is this: to what extent can "the wild" and "the Wilderness" include our own species, and how? Since conundrums by definition are unanswerable, this riddle may be too. But given the extent to which many people care passionately about what they consider *the wild*, governable or ungovernable, this wild riddle is at least worth talking about.

Conundrum Hot Springs is no different from thousands of special places that have become more and more trodden with population and recreational growth. But were it not for the springs' lucky inclusion within a federal wilderness area, they would likely be utterly overrun by crowds come to enjoy them via Jeep and truck and dirt bike, not on foot. Indeed, a couple of watersheds away, around the glorious Cumberland Pass, ATVs ravage the arctic-alpine tundra that lies outside a designated wilderness area. Wilderness designation is often questioned because it excludes humans except as visitors, while what we consider wilderness was often human habitat in earlier eras, to a degree only recently realized. For example, from the 1960s into the twenty-first century, fieldwork by the geomorphologist-turned-archaeologist James Benedict turned up abundant, never-dreamt-of evidence of native occupation of the high country in the Indian Peaks Wilderness. In 1992, Benedict named one of his key papers "Footprints in the Snow: High-Altitude Cultural Ecology of the Colorado Front Range." Surely Conundrum Hot Springs, across the Continental Divide from Benedict's Niwot Ridge study site, was also well used by humans before ourselves: their seasonal, altitudinal migrations must have included medicinal stop-offs at the place of the soothing waters. But those humans, and their pre-ATV level of impact, are long gone. These days I have to ask, isn't it better to toss the distant wild to the wolves, and just occasional humans, than to all the people, all the time, with all their modern engines and appetites?

Happily, it needn't come to such a stark either/or for the great majority of the land. The national forests, by design and slogan, are "the land of many uses." The formal wilderness areas are just one of those many uses, and humans are abundantly accommodated in all the other zones. Nor is the wild entirely absent from them, whether logged, mined, grazed, or drilled. In fact, the wild goes down much deeper than that—all the way to the cracks in the city sidewalks. For here's the big secret: it's not a matter of "wild" or "nonwild." Wildness (in the sense of that which takes us out of ourselves) exists all along a great, big continuum—a sliding scale, graduated not in numbers but in degrees of differentiation from the human quotidian. And why should this surprise us? We have come to learn that many qualities once seen in black-or-white, either/or terms—character, art versus craft, sociopathy, race, beauty and ugliness, and certainly gender—are actually present as continuums. Most dualisms and dichotomies blur upon honest examination.

My own sense of the wild gradient began with a beguiling stump on the corner and traveled from there along an actual, physical continuum—a ditch. But first, that stump: Walking to kindergarten, my mother and I stopped beside it daily to poke for beetles. As long as I was still too small to roam, our backyard contained my multitudes. Soon I graduated to the High Line Canal, an irrigation ditch coursing the altitudinal contours across the landscapes of Greater Denver, carrying Platte River water from its mouth at the edge of the Rockies out onto the plains near the present Denver International Airport. I first met it as a young boy in love with, but too far from, the mountains. Living on the east side of Denver, I might as well have been in Kansas for all the access I had to the Rockies. However, I found that I could escape my raw, suburban tract by traipsing off to the old ditch, and along it, as far as my short legs and the long days would allow. I learned my butterflies there—the wood nymphs and admirals, coppers and skippers—and how mountain species would come down to the plains along the green corridor of the canal, and vice versa. I also learned that the farther west I roamed, beyond the city and toward the foothills, the wilder and more diverse in species things grew. And when eventually the butterflies drew me beyond the hogback and up into the Indian Peaks, I learned what real Wilderness was.

In the same year as the passage of the Wilderness Act (1964), I gained the mobility to reach the high wilds that the act set out to protect. Not long after that, I went to college, read Roderick Nash's *Wilderness and the American Mind*, and understood why I'd felt what I had—and that I would always fight to preserve Wilderness in the world. I also read Leopold, the Muries, and Marshall. Over the next half century, I sought deep wilderness from Mount Bierstadt in the Colorado Front Range to the Brooks Range of Alaska; from the High Sierra to the Himalayas, the Pennines to the Pamir, the Dolomites to the Dark Divide; from the Astrolabe Range of Papua New Guinea to the Qin Ling Mountains of Shaanxi, China, to the Vatnajökull of southeast Iceland, and many other places in between and along the way: all *Wild*, with a capital *W*; some occupied by my own species, some not. And after it all, I still believed (and believe) Thoreau when he wrote, "We can never have enough of nature. We need to witness our own limits transgressed, and some life pasturing freely where we never wander" (306).

Yet I never lost my feel for the wild in its most compressed, contained, and essential forms: the moss and blossom in a sidewalk crack; that ragged old ditch on the backside of Denver; and the marsh-cum-dump on my college campus, which we young conservationists saved along the way while we were also campaigning for North Cascades National Park. When Justice William O. Douglas accepted our invitation to come march with us to protect the Glacier Peak Wilderness Area from Kennecott Copper, it gave me the same kind of thrill as leading a few hundred marchers, with trees in hand, down to the campus landfill to occupy it and declare its destiny to be the habitat it once was, instead of the parking lots it was intended to become. And even as much was being lost on every side, all these things came to pass: the mountain fastness, the urban wetland, and so much more in between—the whole great panoply of wildness, seeking a future among our species' unslakable demands.

Having learned the love of damaged lands from the High Line Canal and the beat-up old fields-becoming-suburbs of Denver's hinterlands, I eventually made my home in the several-times-logged-off lands of the Willapa Hills. And now I've lived more than half my life in a sparsely populated rural county among manhandled forests and fields, finding beauty and, yes, wildness among the clear-cuts of Willapa when I cannot make it up to the Olympic Wilderness in the north. In some ways, I am right back where I started: fascinated by a stump on the corner.

Now that we understand that the general condition of wildness lives along a string that stretches from the back alleys of Gotham to the far peaks of Shangri-La (and I have hiked in a place called just that, in Diqing Tibetan Autonomous Prefecture, Yunnan), we can ask what to call the knots along that elastic string. I don't really worry much about "wild." It's one of those ambidextrous words, like the Irish *craic* or the German *Gemütlichkeit*, that is difficult to define, but you know it when you feel it. "Wildland" goes a bit beyond and indicates places that retain some notable degree of wildness despite their history; I called my book about my old ditch in Colorado *The Thunder Tree: Lessons from an Urban Wildland*, and that's not really an oxymoron. But then there is "wilderness," or even more so, "Wilderness." Here I draw upon the thoughts of others. Thoreau: "Life pasturing freely" (306). Dr. Johnson: "a tract of solitude and savageness" (2278). Bob Marshall: "possesses no possibility of conveyance by any mechanical means and is sufficiently spacious that a person in crossing it must have the experience of sleeping out" (141). Aldo Leopold: "a disclaimer of the biotic arrogance" (6). Or how about this, from Shann Ray's novel *American Copper*: "the wilderness, where she could be alone in great tracts of land, inviolable and fierce of their own accord" (16). Many commentators speak of the wild as being "self-organized." I'm not entirely sure what that really means. But I do understand "inviolable and fierce of their own accord," and it is a way of denoting that far end of the continuum—the Big Wild, the Wilderness.

When I was a graduate student in forestry at the University of Washington, I took a course in wilderness studies from Forest Service scientist John C. Hendee. In one exercise, we took a test that was supposed to measure our position on a scale of "wildernism." As a strong and active advocate for wilderness, I was appalled and embarrassed when I came out as a "weak wildernist." The reason? I had checked that I enjoy driving on small forest roads. Well, I still do . . . very much. Yet I also enjoy walking in the Big Wild, the Deep Wild. And I still don't find the two pleasures to be at odds. They're some distance apart on the continuum, but they both take one into greater contact with the extrahuman. I firmly support road ripping to increase the sizes of roadless areas, and I would always oppose a new road into a designated, or de facto, roadless area. But I would also, always, take that bowered lane ahead in my small, old car with zeal and then, I hope, park it and walk beyond the ruts, tank traps, or gates—walking the wild gradient.

In 1990, I spent much of an autumn in Washington State's largest (55,000 acres) de facto but undesignated wilderness, the Dark Divide, in connection with my book *Where Bigfoot Walks: Crossing the Dark Divide*. The Dark Divide is so chopped about and entered along its edges that the US Forest Service used to refer to it as "Amoeba." The experiences I had convinced me that motorcycles need to be eliminated from its ancient trails, a road or two ripped, and its forests protected beyond the vulnerable provisions of the Clinton Forest Plan and its roadless rules. That can be accomplished only by inclusion in the national wilderness protection system. However, the Third Congressional District, through gerrymandering, has become a virtual sinecure for the party that never originates, seldom supports, and often blocks new wilderness areas.

Apart from obvious commercial considerations, what is it about wilderness opponents, such as our congresswoman, that so engages their animus? Very often it has to do with the exclusion of human uses (such as dirt bikes, mountain bikes, and off-road vehicles) from wilderness areas. In many a wilderness hearing, discussion, or reading, I have heard the charge that wilderness protection is somehow "elitist." This goes back to at least 1926, when forester Howard W. Flint, in an article titled "Wasted Wilderness," attacked Aldo Leopold's wilderness advocacy as being for the "elect few." Many writers since, including Peter Kahn, Baird Callicott, and Peter Sutter, have considered the "elitist" charge at length and have found it to be generally self-interested. Maybe there have been instances when a true elite has employed the idea of wilderness in its favor—such as when the Rockefellers bought up Jackson Hole to protect their neighborhood—but would anyone today wish to give up their subsequent gift to the nation, Grand Teton National Park?

One unfortunate legacy of wilderness debate in recent years has been to render "wild" just a four-letter word in some people's vocabularies (many of them in Congress) and to give sanction to the redemonizing of Wilderness. Once demonized by the superstitious who were simply afraid of the sublime wild (sometimes with good reason), wilderness was later largely tamed, then valued, cherished, embraced as a vital legacy, and protected in its lingering margins by the force of law, only to be damned all over again as elitist! That won't do. We knew better, even as dumb college kids: anyone with a pair of boots or sneakers and a crappy knapsack and a canteen knew that he or she could head to the hills with friends, by bus or hitchhiking if necessary, walk into a federal wilderness area, and keep walking—for *free*. The wilderness, to the young conservationists with whom I came of age, was—and still is—the antithesis of elitist, unless that "elite" included all the plants and animals that lived there, along with every cash-strapped kid or working stiff who could put one foot in front of the other on a trail. I do not know a more democratic ideal than that of Wilderness sensu 1964.

Now, regarding human beings—the upright primate species designated *Homo sapiens* ("all alike and intelligent"—my, what an optimist Linnaeus really was! Let us hope we someday deserve the name)—in or out of the wild, let's get a couple of things straight. First, we are just another species of primate, and everything we manufacture or perform is a product of an evolved species of upright hominoid ape. Ergo, all of us and all of our productions are part of nature, if nature is everything, which in my view it patently is. Steinbeck, in *Log from the Sea of Cortez*, said, "Most of the feeling we call religious . . . is an attempt to say that man is related to the whole thing... all things are one thing, and that one thing is all things" (257). Or as Nirvana, in their song "All Apologies," more succinctly put it, "All in all is all we are."

So far, we're good: the idea that there is some essential border between people and the rest of nature is one of the most dangerous dualisms in the world, lending license to all manner of enormities against other species and the land that supports them. No one who understands that humans and nature are indivisible, parts of one continuum, could sanction the Alberta tar sand crimes against the taiga and its people or mountaintop removal in West Virginia. If people are part of nature and the wild lies "out there" in nature, then the wild also dwells within, as Gary Snyder has shown us. Thoreau said, "We are conscious of an animal within us" (210). And Steinbeck asked, "Why do we so dread to think of our species as a species, our eyes the nebulae, universes in our cells?" (314). Why indeed? Antipathy to Wilderness must often be rooted in this question.

Here's another fact that might help us to weasel in a little deeper: Wilderness, sensu stricto, existed for billions of years prior to the advent of humans, and it will exist forever, at least until earth's molecules are assimilated elsewhere, and then it will continue to exist somewhere else. Humans have coexisted with the wild (and been part of it, and vice versa) for only the tiniest fraction of that time—maybe 1/5,000 of it so far. Further, the time will come when we humans will be *extinct in the wild*, for that fate befalls all species—that is, unless technology conjures a means of extinction avoidance, in which case we might actually move beyond biology, maybe *extant* in some cyborg form, but *extinct in the wild* nonetheless. There is no telling *when* we might go extinct, although sensible bets among anyone conscious of natural limits to growth might well wager on sooner rather than later, given our present proclivities. But however long we manage to forestall the inevitable, the fact remains that the wild was here for a long time before us and will continue for a long time after us. That basic understanding tends to place many of our views as wild thinkers in a rather different light and should encourage us to take a longer view of our actions.

The thing is, we are not constrained to love only one part of the relative wild, to hold fast to just one concept of what we should talk about when we talk about wildness. Nor do we need to oppose our conceptions against one another for ideological, intellectual, or political reasons. The world is big and great enough to have and to hold all faces of the relative wild—but only if we continue to work as the activists of old, to allow the world to do so, in the face of boundless human cupidity and insupportable excess.

Why should we work with one hand tied behind our backs, one arrow in our quiver? Can we not have the ever-present wild and Wilderness too? Is it beyond our powers at this late date to maintain the entire blessed continuum?

Which brings us back to people on the land. Heaven knows that outrages have taken place against indigenous or latter-day native people in the name of wildland conservation. Examples are rife on the world scene and in North America's history. It can still happen today. It is also patently clear that some forms of traditional occupation of the land can foster (not create, but support) native plants, animals, and the community in concert with human use. One such example is the acequia culture of the desert Southwest. This is not a mystery: water concentrates life in arid lands, including old, human-dug, well-managed watercourses no less (and sometimes more) than rivers and streams. I knew this from an early age by becoming a near-full-time denizen of a prairie ditch myself and studying its butterflies and plants. Now that the farms are gone and Denver Water has shut all the head gates, dewatering the High Line Canal, the butterflies are in trouble, along with the old cottonwoods and the people who love them. And when the acequias and their keepers are ejected from their valleys, the result can be not only cultural genocide but also a net loss of natural diversity. Gary Nabhan, in *The Desert Smells like Rain*, heartbreakingly documents such a case—the removal of Tohono O'odham from their oasis and the acequias at Quitobaquito Springs when Organ Pipe National Monument was established.

When my own ditch studies led me to England to learn more about butterfly conservation, my eyes opened wide to learn how British orchids and insects, primroses and bluebells have adapted to ancient forms of grassland and woodland management and languish without it. And after common market and Thatcher-era "improvements" had done their worst, my colleagues were finally able to say with documented certainty that *every instance of butterfly endangerment in the United Kingdom can be directly attributed to disruption of traditional agriculture*. At least in the case of the extinct British large blue butterfly, reestablishment of the old ways of grassland management has allowed the butterfly (reintroduced from Sweden) to thrive once again.

Helen Macdonald writes of another British reintroduction, the goshawk, in her book *H Is for Hawk*: "Their existence gives the lie to the thought that the wild is always something untouched by human hearts and hands. The wild can be human work" (8). Surely we should fight for agriculture that returns to such a mutualism, even as we fight to expand wilderness areas and corridors.

So, can the wild persist, is the wild sometimes even *enhanced*, under human occupation? Of course. And that's not mysterious either: human farmers, cattle, and sheep coevolved with native plants and animals in the British Isles since day one after glacial retreat for more than ten thousand years. Acequia culture in the Southwest may be just as old, or older. And there are endless other examples of humans coexisting and coevolving with the rest of the wildland community in a manner that seems beneficial, or at least largely benign, for both. There are notable instances where National Park Service tolerance of established uses seems to work, as I have seen in the Wrangell Mountains of Alaska.

Yet when old, sustainable forms of land and water use are trashed by authority, the result can be socially repugnant and personally tragic. For example, it is the present policy of both Washington and Oregon fish and wildlife commissions and governors to eliminate traditional gillnetting from the mainstem Columbia River on the pretense of conservation, but many believe this is for greater sport fishing revenue. In Wahkiakum County alone, where I live, this dictum will harshly affect at least thirty families, in a county of just four thousand souls. Sometimes the native plants and animals suffer from policies that come down in the name of saving the Wild, as at Quitobaquito. As a former Nature Conservancy land steward, I have felt a similar sense of transgression (if on a smaller scale) when the Nature Conservancy takes down tree forts, sometimes used by generations of children for getting *out there*.

When human uses respect and sweeten the wild, let's recognize and celebrate them and preserve those uses when we can. But let's acknowledge the opposite, too: Pleistocene overkill of North American megafauna, the cedars of Lebanon, Steller's Sea Cow, the Great Auk and the Dodo. Sometimes humans, ancient and modern, have done anything but enhance the integrity of the wild. Or what about when traditional human occupation has long been absent from a wildland and present (or threatened) use—mining, logging, off-road vehicles, shooting, any number of other modern manifestations and infestations—is far from consistent with the well-being of native ecosystems and their constituent species? Protecting wildlands from abuse has much more often been the outcome of wilderness designation than kicking out sustainable users. Just because we are capable of recognizing the injustices that sometimes occur in the establishment of protected areas cannot mean that we must also forswear the tool of strict land zoning, especially when such legal measures are what the facts indicate and what our consciences require. Sometimes you need only a hoe; sometimes, a hammer.

In the spring of 2014, I attended a wonderful party thrown by the Polly Dyer Cascadia Broadband of the Great Old Broads for Wilderness to celebrate the fiftieth anniversary of the passage of the Wilderness Act. The Broads brought in poets to speak the liturgy. Bill Yake, dressed in period garb, spoke the words of Thoreau in character: "In wildness is the preservation . . ." I read six poems of early wilderness encounter, all describing incidents in the 1960s, the decade of the act's passage. And Clem Stark read the essential text of the Wilderness Act itself, crafted largely by Howard Zahniser and perhaps the most beautifully written measure in American legislative history. It spoke of saving for all Americans, and all people, the kind of land "where the earth and its community of life are untrammeled by man, where man himself is a visitor who does not remain." Or, as Thoreau had it, where we can "witness our own limits transgressed, and some life pasturing freely where we never wander." Or seldom wander, but where we earnestly hope we someday may. And what is wrong with that?

Was the act blindered to the people who had once been there and sometimes still were? Perhaps, but not entirely. Hunting is allowed in wilderness areas, as is grazing—if they are outside national parks, as most of them are. But had the Wilderness Act and its proponents not been as single-minded as they were, that remarkable moment would have passed, Aldo Leopold's roadless area groundwork might all have been wasted, and many of our finest wilderness areas might well have been rendered entirely unsuitable for the purpose by now. Just scan the horizon of gas, oil, and coal development across the West today to get a sense of my meaning. Are there still abuses? Sure—the very difficult task of managing the blessed public lands in a bloated mercantile country such as ours is bound to lead to some bad decisions. But these are no reasons not to celebrate one of the most forward-looking laws in the world annals of conservation.

However overused, acid rained, climate changed, or otherwise compromised from its precontact condition, the wilderness areas will continue to define one end of that great continuum—the Relative Wild. Many other points along that sliding scale may easily be found by anyone who sets foot out the door, in the mud, off the path. Each of us will continue to express our love, concern, and energy toward the varieties and expressions of wildness with which we most identify. And I suspect that for most of us, that will mean more than one, maybe many, notches on the wild dial. I know that for my part, my tastes and loyalties range over the whole spectrum, from end to end: I shall go on loving my local stumps and sidewalk cracks, the moss-rich masonry of a city wall, my dried-up and ragged old ditch, the barbered timberlands that support my human community, and my grandson's backyard, even as I long for, celebrate, work to protect, and sometimes even visit actual Wilderness.

I will also continue to haul my sorry old ass up into the more and less wild parts of the Gifford Pinchot National Forest, both by car and by foot, where I am welcome to swing my butterfly net and learn new things, every chance I get. The Klickitats are long gone from the Indian Race Track in Indian Heaven Wilderness Area, but several tribes continue to set up berry camps in autumn just to the north. I honor their once and future lifeways when I venture up there, by harvesting huckleberries only in the area designated for whites in the famous 1932 Handshake Agreement between the Yakamas and the US Forest Service. And my keenest devotion as a lifelong conservationist remains the eventual establishment of a Dark Divide Wilderness Area to protect the heart of the land where Bigfoot walks from the dirt bikes that bedevil its trails and meadows today and from the threat of the chainsaw, the dam, the mine, the road—the whole catastrophe, in Zorba's words.

Then, if ever as a people, as a world, we get past the catastrophe, at least there will still be a few places that we did not choose to waste along the way. After all, maybe our utterly unsustainable capitalist "economy" will somehow moderate such as to allow one or another vision of working with the wild to take and we will enter a brighter time, one with greater capacity and amplitude for coexistence among ourselves and all the other species. If so, we will be glad for the options we have preserved by setting aside some slices of Wilderness. And if not? Well, then at least we will have preserved the best possible building blocks for evolution's next iteration: banks of genetic diversity that give natural selection another chance. This is why I say the Wilderness Act is so forward looking: because of all the laws by which we live, it is almost alone in looking to the good it can do even beyond human existence.

Sometimes I muse on the phrase "Extinct in the Wild." From an early age, I found that phrase heartrending. I remember reading it in one of the nature books that made up most of my earliest diet as a reader and how I recoiled from its implications. *"Extinct in the wild?"* I worked it out and understood what it meant—that I could never, ever, go *out there* and find it, whatever it was. So, I suppose, the word "wild" and the idea of "out there" came to mean one and the same thing to me. To a large degree, they still do.

On a recent trip to the Scottish Highlands, my friend and I came to a place called Borgie Glen. Standing there, gazing off into the high purple heather hills, we noticed a small sign pointing toward a dark forest of Scots pine, deep in a receding crease of the Cairngorms: "The Unknown," it read. And there I realized—*that's* what I most want out of this slippery word "wild." I don't want to live in a world become so relative that The Unknown itself is Extinct in the Wild.

## *References*

Benedict, James B. "Footprints in the Snow: High-Altitude Cultural Ecology of the Colorado Front Range, USA." *Arctic and Alpine Research* 24, no. 1 (1992): 1–16.

Flint, Howard. "Wasted Wilderness." *American Forests and Forest Life* 32, no. 7 (1926): 407–10.

Johnson, Samuel. *A Dictionary of the English Language: A Digital Edition of the 1755 Classic by Samuel Johnson.* Edited by Brandi Besalke. http://johnsonsdictionaryonline.com/?page_id=7070&i=2278.

Leopold, Aldo. "Why the Wilderness Society?" *The Living Wilderness* 1, no. 1 (September 1935): 6.

Macdonald, Helen. *H Is for Hawk.* New York: Grove Press, 2014.

Marshall, Robert. "The Problem of the Wilderness." *Scientific Monthly* 30, no. 2 (1930): 141–48.

Nabhan, Gary Paul. *The Desert Smells like Rain: A Naturalist in O'odham Country.* Tucson: University of Arizona Press, 2002.

Nash, Roderick Frazier. *Wilderness and the American Mind.* 5th ed. New Haven, CT: Yale University Press, 2014.

Pyle, Robert Michael. *The Thunder Tree: Lessons from an Urban Wildland.* New York: Houghton Mifflin, 1993.

———. *Where Bigfoot Walks: Crossing the Dark Divide.* New York: Houghton Mifflin, 1997.

Ray, Shann. *American Copper.* Lakewood, CO: Unbridled Books, 2015.

Steinbeck, John. *The Log from the Sea of Cortez.* New York: Penguin Classics, 1986.

Thoreau, Henry D. *Walden.* 1854. Edited with annotations by Jeffrey S. Cramer. New Haven, CT: Yale University Press, 2004.

# 3

## NO WORD

*Enrique Salmón*

There is no word for the concept *wild* in my native language of Rarámuri. The Rarámuri are a large population of people indigenous to the Sierra Madre mountains of Chihuahua, Mexico. The region is also known as the Sierra Tarahumara. There are roughly seventy thousand Rarámuri who continue to speak our language and live a subsistence lifestyle in this extremely remote, rugged, and biologically diverse landscape.

The Sierra Tarahumara comprises eight physiognomic vegetation types, including montane evergreen and oak-coniferous woodlands, tropical deciduous forests, oak savannas, chaparral, shortgrass prairie, subtropical thornscrub, and subtropical desert fringe. This area houses the third-largest concentration of biodiversity in the world with four thousand vascular plant species, one hundred fifty of which are endemic (Ramamoorthy et al. 1993, 44–55). In addition, the Rarámuri people have managed to hybridize eighteen pre-Columbian food crops, including agaves, pepperweed, panic grass, tepary beans, squashes, and several maize varieties. It is no accident that the area is so diverse. There is a strong link between cultural diversity and biodiversity. Indigenous land managers have practiced a sustainable lifestyle for several centuries. In other words, the people of the Sierra Tarahumara have acted as direct keystone species among the ecosystems and habitats of this region.

For the longest time, the conservation and environmental movement assumed that the human-environment interface always resulted in negative outcomes for the land. The last two hundred years of exponential human population growth coupled with mass expansions of industry and globalization has certainly done little to balance out the equation. As a result, until recently, researchers had not considered the possibility that humans could actually enhance their landscapes, that human communities might actually play a role in increasing diversity, or that humans are as essential to the ecological functioning of a landscape as saguaro cacti are to the Sonoran Desert. Therefore I suggest that human communities could be a keystone species in some ecological systems.

The Great Yellowstone Fire of 1988 was a catalyst for waking up conservationists to the once taboo concept of allowing wildfires to continue unabated. After the fire, historical studies emerged showing that forest fuel levels, reflected in the amount of understory and relative thickness of new and old tree growth, were at their highest in recent decades. Land managers began to question what, in the past, allowed for conditions where fuel levels were lower, thereby creating conditions for less-intense fires that were not as catastrophic as the ones burning today. The answers that slowly emerged all pointed to pre-Columbian human manipulation of these landscapes.

Wherever people have sustainably cultivated a landscape, there has emerged a culturally recognized and sanctioned pattern of using and talking about and interacting with that environment. It should not be surprising to notice, then, that residing in a majority of the most biodiverse regions of the world are human communities that continue their cultural legacies, a fact that is reflected first in the survival of their language and worldviews.

A Rarámuri worldview does not differentiate or separate ontological spaces beyond and between the human and nonhuman worlds. We feel that we are directly related to everything around us. The trees are us; we are the trees. I am rain; rain is me. The rain is all around me; it aligns inside me. These feelings are reflected in how the Rarámuri classify things such as plants.

To the Rarámuri, this classification system includes gendered plants, plants identified as *Chabochi* (white people), plants who are indigenous, and plants who are Rarámuri. The categories of plants are indistinguishable from the Rarámuri categories of humans and, to a large extent, mirror human social categories. In addition, the moral and behavioral attributes of these human social categories are projected onto the corresponding categories of plant people. I think of this interconnected view of all life as a kincentric ecology, in which everything is a relative. Such a worldview and corresponding classification system of the natural world leaves no room or need for a category such as the notion of "wild." If everything is a relative, and if we are interconnected to everything around us, then there is no need for a category of thought nor a specific term to denote a part of the world that is separate from ours.

To a Rarámuri, there are certainly parts and facets of the natural world that are dangerous. There are plants and animals that one has to be careful around and have deep knowledge of before engaging with them. Take, for example, the plants categorized as *Chabochi*. *Chabochis* are nonindigenous humans who are not Rarámuri. Generally, the word is used to identify white people. *Chabochi* means literally "the bearded ones." It is said that Diablo has a beard. In addition, *Chabochi* are considered "children of Diablo." To Rarámuri, *Chabochis* are mean and stingy and fight all the time. They are dangerous and cannot be trusted. In this model of the universe, good and evil are at odds. Diablo constantly threatens the Rarámuri while they assiduously perform the proper rituals to keep *Onorúame* (our Creator) strong for fighting Diablo's wickedness.

It may not come as a surprise that the plants in this category are hallucinogenic and dangerous. *Uirí* (*Datura sp.*) is *Chabochi*. Other plants who are *Chabochi* include bakánawi and uchuri. Bakánawi is a group of three plants including *Scirpus acutus*, *Coryphantha compacta*, and *Ipomoea purpurea*. Depending on who is being asked, bakánawi are not always *Chabochi* and are sometimes Rarámuri. Uchuri is another group of three kinds of cacti. There are some types of híkuli (peyote) who are *Chabochis*, including híkuli mulatto and híkuli sonami. Then there is Oliarki (*Ricinus communis*), which is a highly toxic plant. It is a *Chabochi* as well. It is noteworthy that *Ricinus communis* was introduced to the region by Europeans.

The Rarámuri schema of human-plant categories includes indigenous groups. Although some people consider bakánawi to be *Chabochi*, some feel that bakánawi were also Apaches. This was due to their fierceness and unpredictability. Some híkuli are Apaches as well because they are so dangerous. It is not surprising that the Apaches are singled out as related to dangerous plants. The Rarámuri and other indigenous groups in Northwest Mexico recall an unpleasant association with the Apaches. The Apaches often raided stores of corn and animals. They stole children and women and killed Rarámuri warriors. The animosity between the two peoples revealed itself when Rarámuri warriors volunteered to scout for the Mexican army during the 1880s, when Apache raids were especially virulent. It is reported by the US Indian Service that a Rarámuri named Mauricio single-handedly killed the war chief Victorio in 1880.

Up to this point, I have been focused on my native language. However, in my nearly four decades of speaking with, hanging around, and sharing notes with Indigenous peoples, I have yet to identify a Native language that includes the term *wild* in its nomenclature. Since the beginning of European colonization, most indigenous languages have borrowed and added English, Spanish, and French terminology to their vocabularies. The Rarámuri language, for example, includes the term *cimarron*. Spanish speakers will recognize the word as a descriptor for people who are wild, untamed, rough, uncouth, and even lazy. It was also used by early Caribbean colonizers to identify escaped slaves. Among the Rarámuri, the word is an identifier for those who have not allowed themselves to become baptized. Unbaptized Rarámuri are considered uncivilized. Baptized Rarámuri act wary and a bit afraid around the *cimarronis*.

One warm afternoon down along the edge of the Barranca de Cobre in Chihuahua, Mexico, I was sipping *cokas* (Pepsis) with some Rarámuri friends. We were in the small one-room *Conasupo* (government-sponsored general store and trading post) at the edge of the community of Norogachi. In walked three traditionally dressed Rarámuri. Wrapped around their heads were thickly rolled calico bandanas that swung low to the middle of their backs. The sleeves on their white tunics were billowy bloused, and they sported breechclouts that were embroidered with flower, river, and mountain motifs. I noticed they were leaner and darker than the Rarámuri around Norogachi. Their dress was of a slightly different cut, and they wore leather-soled sandals instead of the more commonly seen car-tire sandals. I also noticed that they carried water gourds and bows and arrows. When they entered, everyone quickly became quiet and cautiously stared at the newcomers. The three asked for *cokas*, which they quickly and quietly sipped. Their eyes darted around at the contents of the room and at us. We must have looked like the timid rabbits that they, no doubt, hunted with their bows. One of them traded some herbs for some colorful cloth. They finished their drinks and quickly left. No one spoke for a moment, then I broke the silence and asked who those guys were. Someone whispered, "Cimarronis."

Although the word *wild* does not exist in the Rarámuri or other American Indian vocabularies, I can appreciate the concept of wild and wildness in its current context. Modern industrialized humans have become so far removed from the natural world and our interrelationship with it that we need a mental and linguistic way to remind ourselves of our primal past when we were not separate from, but totally immersed in, our natural surroundings. I believe this explains the social, cultural, and political drive for the development of national parks, national forests, and backpacking. It explains why people will hang scenes of Yosemite granite, Yellowstone waterfalls, Alaskan mountains, and Arizona desertscapes on their office walls sixty floors above large city centers. It also explains why modern Americans so revere the idea of the noble, stoic, and proud American Indian, even claiming that somewhere in their hazy family lineage a great-great aunt, grandfather, third cousin twice removed, or some distant ancestor was an indigenous person. It is as if knowing that there are seemingly untouched swatches of landscapes and that their genes include some connection to the natural world makes them more fully human.

This actually might be the case. For our two hundred thousand years as modern *Homo sapiens*, we have lived in what many would refer to as a natural state. Everything we required in the form of food, clothing, and tools was derived and/or constructed from resources provided from our immediate surroundings. In many cases, human populations figured out how to live sustainably with their local ecosystems and became keystone species within their local environments. Only very recently—perhaps in a little more than one hundred years, coinciding with the advent of fossil fuel-powered machines—have humans relied less on nature and more on technology to unnaturally manipulate and in many cases harm their natural surroundings.

This technological shift and the separation it created may be the most prominent reasons that *wild* means different things to different people and populations. Among Euro-American, English-speaking, Westernized people, *wild* seems to refer to things that are undisciplined, uncultivated, crazy, untamed, feral, disorderly, unruly, perhaps poor—generally, things outside the realm of the "civilized."

I once asked a group of my college students to react, using only a single word or a couple of words, to the word *wild*. Their responses included Tarzan, animal, nature, prairies, ghetto, natural disaster, untamed, undomesticated, plants, freedom, Africa on television, jungle, and *Planet of the Apes*. I was not surprised by such a range of reactions. Notions about the term *wild* are extremely contextual and used to convey different meanings for differing reasons. I am not equipped to delve deeply into why this is the case. However, I can offer some ideas about *wild* and its meanings—and lack of meanings—from a specific cultural perspective.

I believe that humans are genetically predisposed to a more nature-based lifestyle than those in which most people currently engage. E. O. Wilson proposed a similar concept in his book *Biophilia*. Our bodies, moods, and even brain chemicals are directly influenced by lunar shifts, climatic changes, and even magnetic shifts emanating from deep within the earth. In addition, we still depend on the natural world for our oxygen, food, and water. However, the portions of our brains used for reasoning developed rapidly and began to create human-centered linguistic categories in an attempt to make more sense of the universe. An effect has been for people to mentally categorize humans as separate from everything else. It seems that *Homo modernus industrialii* falsely constructed a set of words, and therefore a worldview, that hierarchically sets themselves above and apart from everything else.

The concept of wild, then, frees modern individuals to place themselves beyond any socially constructed diorama where humans are not just another group of very intelligent animals. The notion of *wild* permits us to justify actions of ours that are systematically harming the very thing that is keeping us alive. We can develop wilderness areas; capture parts of the so-called wilderness for study in museums, labs, and zoos; and even go "back into the wild," at least temporarily. Wild, wilderness, and wildness are many things, but they are not natural and certainly not real.

In 1492, when the ocean was still blue, as the childhood rhyme informs us, the first wave of Europeans found themselves on this new continent. They also came into contact with groups of humans that did not quite meet their notions of humanness. My ancestors and others like mine did not share the mental categories that placed humans as separate from and above everything else. Many Europeans considered American Indians a kind of subspecies of human. We quickly became the Calibans of the known world. In our persons, often described as savage or beastly, the European gaze saw only extensions of their concept of a new wilderness. All forms of American Indian keystone species land management went unrecognized. Our very sophisticated agricultural practices were deemed regressive or backward, and our uniquely stratified and complex social systems were placed in the category of devil worship. The word *wild* was often used as one of several degrading adjectives when referring to us. Therefore, in the minds of those first European colonists, the newly identified "wild Indians" were a sociolinguistic construction that had real consequences.

One of the most important facets of the human mind is that it can create reality. We do this when we name things. Naming affords a thing form, substance, and matter. It makes it real. We are mentally able to name a concrete reality such as a rock, then employ linguistic metaphors that blend cultural concepts together to form new mental associations that are just as real. We even do this with abstract concepts such as love and death. These new mental spaces become culturally shared models that are based on multidimensional projections. Let's call this *cognitive blending*.

One can think of the concept of cognitive blending as similar to preparing a pot of posole. It is possible to gather together the necessary ingredients for posole but still not have the food. From one mental input can come the pot and water, and from the other the dry ingredients. But what makes them blend together as posole is the care given during the preparation by one's mother or grandmother and the years of preparing it a certain way, as her grandmother did. It is essential to know exactly when to stop boiling the corn and just how much spice to pour into the pot. These and other essential elements that make up the ceremony of preparing posole cannot be derived from the inputs; they are part of the unique structure of the blend that creates the meal. *Blended spaces* are similar.

Among my people, it is understood that corn plants are females. In this metaphor, the mental domain is "female children" and the concrete domain is "corn plants." The attributes of female children are projected onto growing corn plants. In the middle space, female children and corn plants share the abstract elements of caretaking, stewardship, and healthy growth.

In the blended space, prelinguistic concepts unattainable from the other inputs become apparent and crucial for a holistic understanding of the metaphor. Here we find that the corn plants are afforded animus. Like female children, they have emotions and needs. They must be nurtured, cared for, given attention, prayed for, and protected. An interdependence exists between the corn plants and humans that is also found between parents and female children. The metaphor implies an emotional connection as well, between the speaker and the corn. In return, the corn and children provide nourishment to both the caretaker and the culture. The result is that the relatively abstract domain of corn plants with animus has been conceptualized in terms of the more concrete one of female children. In addition, the underlying Rarámuri metaphor that plants are people, and therefore kin, lends further conceptual connections to female corn plants. The interdependence that exists between the small Rarámuri communities for survival extends to their plant relatives and kin who contribute to the survival of the people. Mental blended spaces often inherit additional meanings from social context, cultural frames, and background. Richer structure comes from these additions, which become the locus of cultural truth. The point is that if humans can mentally create the reality of *wild* and *wildness*, then perhaps we can also alter, reclaim, or reinvent their meanings.

At one point in American linguistic history, the word *environment* was used when referring to one's immediate surroundings or to refer to natural areas inhabited by nondomesticated animals or plants. However, it seems today that when speaking of *the environment*, most people are extending the original meaning of the word, making reference to nonurbanized ecosystems—air, natural rivers, lakes, ocean systems, forests, deserts, and mountains. We have altered the meaning and the grammatical use of a term to fit today's needs and culturally shared mental spaces. The word has been transformed from a generalized noun to one that is more specific.

It is commonly said: We, us humans, have to protect *the environment*; We must take care not to pollute *the environment*; *The environment* can provide for us if we treat it right; or It is so nice to get into *the environment*. *The environment* has metaphorically assumed entity-like status in our language and, therefore, our minds. These examples offer evidence of how the meaning and use of a word connected to nature can be changed. Of course, in this case I am only referring to American English speakers. Back in Rarámuri country, there is no word equal to that of *environment*.

We may be stuck with the word *wild*. However, like the use of the word *environment*, its culturally constructed meanings might be reshaped toward something that celebrates direct human connections to their natural surroundings. Most modern industrialized humans maintain an indirect relationship with their natural surroundings. Take as an example our relationship with our food. Only 2 percent of Americans are farmers. Around 20 percent of Americans raise some food in their home gardens, but 86 percent of those foods tend to be tomatoes. Therefore, the majority of food that Americans eat is transported, usually from considerable distances, before it reaches our local grocery stores. When asked if they like guacamole, most of my students say *yes*. When asked what the main ingredient in guacamole is, most still seem to know that it is avocado. However, when asked how avocados grow, most have no clue that they come from trees in warmer climates. This is an indirect relationship with a part of nature that feeds us.

As another example, in 2014, less than 1 percent of Americans had ever been backpacking. That means that more than 99 percent of Americans have never intentionally experienced sleeping out under a roofless sky. They have never drunk water from a spring or stream, have never cooked over an open fire or a tiny gas stove. Most have never seen an untamed animal outside of a zoo, and most have never had the realization that "Out here in the woods, I am not at the top of the food chain." With these two examples, it becomes apparent why the notion of *wild* is a part of modern nomenclature; our modern worldview and lifestyles continually reemphasize our disconnection from the natural world. What if the wild began to mean something else?

In the Rarámuri language, there is a way to refer to the land. The word is *gawi wachi*. It translates into something like "that which nurtures us." Among the Navajo, they refer to their homeland with the word *dinetah*, which roughly means "it is among us." Even traditional Hispanos in northern New Mexico and southern Colorado make reference to their deep connection to their lands with the word *querencia*, which means "the place that holds my heart." What if *wild* then was reinvented to mean something that stirs that primal part of modern humans to recognize that, yes, I am a *Homo modernus industrialii*, and I share breath with the trees that are in and around my neighborhood, and the avocado in my guacamole is a relative.

I realize my musings to be mere fantasies. What is more realistic is that soon modern humans will begin to recognize that we are one of several keystone species within the ecological realms that we occupy. We are crucial to the success or failure of our ecosystems. Unfortunately, there is no word in English to describe these complex understandings.

## *References*

Ramamoorthy, T. P., Robert Bye, Antonio Lot, and John Fa. *Biological Diversity of Mexico: Origins and Distribution*. Oxford: Oxford University Press, 1993.

Wilson, E. O. *Biophilia*. Cambridge, MA: Harvard University Press, 1984.

# 4

## THE EDGE OF ANOMALY

*Curt Meine*

Wilderness exists in all degrees, from the little accidental wild spot at the head of a ravine in a Corn Belt woodlot to vast expanses of virgin country. . . . Wilderness is a relative condition. —Aldo Leopold, "Wilderness as a Form of Land Use" (1925, 399)

Our four-car caravan leaves the creekside parking lot, winds across the valley, follows the curves of a narrow road, climbs up a bank of steep oak-forested hills, and rolls through open pastures and upland crop fields toward the ridgetop farm of Joseph Haugen. Most of the twenty university students have no idea where we are. There is no reason they should. Some have come from ten thousand miles away. But here we are, a few miles from the Mississippi River in western Wisconsin, heading to a small farm where eighty-eight-year-old Joseph now lives alone. His older brother Ernest, with whom Joseph lived and farmed this land all their lives, died two years ago. The bachelor brothers had sold their dairy herd some years before but kept three Jersey cows for milk, for company, and for continuity's sake. The lede of the obituary in the *La Crosse Tribune* read, "Ernest Haugen, a farmer and champion of land conservation, died Thursday at age 90, just five weeks after milking his last cow."

We pull into the Haugen farm. Its modest old farmhouse sits on one side of the road, outbuildings on the other—a not uncommon arrangement in this part of Wisconsin. The students settle onto the sloping lawn, sip from their water bottles, and enjoy the midmorning summer sun. My friend Jon Lee, who farms nearby, and I go to fetch Joseph and three wooden chairs. Jon knocks on the door. "Good morning, Joseph!" Jon says loudly. Joseph's hearing has faded. "Hello, Yon." Joseph speaks with a second-generation Norwegian accent. He is slight and wiry. He's a smiler and lights up as we meet him at the door. He wears blue jeans, a plaid short-sleeved shirt, and a ball cap and walks haltingly with his cane across the grass. For someone who comes close to fitting the stereotype of the stoic Norwegian bachelor farmer, he enjoys our having come for a visit. But he also tires more easily these days, so we must make good use of the time with him.

The Haugens and their neighbors in the Coon Creek watershed were revolutionaries. In the mid-1930s, the farmers of Coon Valley came to terms with their land, with each other, and with the actions of their own forebears. Three generations of postsettlement farming and flooding had ravaged the watersheds of these steep-walled valleys. The region's fine loess soils, rendered vulnerable by heavy grazing, constant mono-cropping, and up- and downhill plowing, melted away in heavy rains. Chasmic gullies ate into the hillsides, the eroded soils burying downstream homes, farms, and businesses. Down-valley, in Chaseburg, you can find the tops of chimneys just barely poking out above the ground. In 1935, conservationist Aldo Leopold summarized the situation: "Coon Valley is one of a thousand farm communities, which through the abuse of its originally rich soil, has not only filled the national dinner pail, but has created the Mississippi flood problem, the navigation problem, the overproduction problem, and the problem of its own future continuity" (206–7).

The solution came in the form of a watershed-wide community response, as more than four hundred farmers broke with past practice and adopted novel soil and water conservation methods. The techniques were in many cases experimental, the aims basic. Keep the soil in place. Slow the water. Start at the upslope sources. Moderate the water's infiltration. Adjust land use to fit the degree of slope. Plow and crop along the natural contours of the land. Intersperse and rotate crops. Take the cows off the steep slopes. Repair and revegetate the stream banks. Plant food and cover for wildlife. To make all this happen, work with the newly established Soil Conservation Service, university specialists (including Leopold), the callow boys of the Civilian Conservation Corps, the town bankers, and local governments.

And don't just do all this for a season, or a year, or until the soil erosion crisis passes, or as long as the government funding lasts. Commit to it. For a lifetime. Or longer.

Beyond the technical details of the innovative field projects, the work at Coon Valley and throughout the region reflected a radical new approach to conservation. Here, conservation focused not on protecting large expanses of public land but on the restoration of private lands and collaboration among private landowners. It did not treat parcels of land in isolation but involved an entire community and a whole watershed. It brought in specialized agronomists and soil scientists and foresters and wildlife biologists, but it integrated their perspectives, skills, and expertise in the field. It recognized the need to rebuild and sustain the economic productivity of the land but saw that this could only be achieved by recognizing a broader set of values and respecting the native qualities and wild ways of the land itself. In repairing the Coon Valley landscape, its people helped redefine the very meaning of conservation.

Joseph is among the last living links to that generation of revolutionaries. That is why we wanted the students to meet him.

We sit in our chairs on the lawn and share a few stories and questions. Joseph, in his lilting accent, recounts how all this looked eighty years before. He remembers the government engineers who helped lay out the contours. He describes the changes in wildlife: fewer grouse and quail now, but the return of deer, turkey, even rabbits. He recalls life on the farm with his brother. There is a famous story told about Joseph. He is said to hold the all-time record for sustained milking, having attended to his cows every single day at 5:00 a.m. and 5:00 p.m. for forty-seven straight years. In all that time, he missed only one milking. And that was because he was called into town—"for y-ury duty!" Before Ernest died, the Haugen brothers put their names to a conservation easement for their 160-acre farm, protecting it forever from development. Their final act of grace. The revolution continues.

Our Q&A session comes to a close. The students have sat rapt for the last half hour, half bemused, half in awe. They gather themselves together and we prepare to head on to the next stop on our tour. We help Joseph back into his house. The mudroom is dusty and weathered. There is kindling in a weathered wood box. Joseph heats and cooks with a wood stove. He smiles as we say good-bye.

Joseph Haugen is an anomaly.

\* \* \*

I live a hundred miles east of Joseph, on the other edge of anomaly.

We both dwell in the Driftless Area, where the flat land wrinkles. Where the back roads and corn rows are not straight but curve around tight bends and through sweeping arcs. Where land uses don't follow the checkerboard grid of the land surveyors' township and range lines—the rigid pattern familiar to anyone who has flown over the American Midwest—but go awry and get twisted. Where reality loosens the fixed grip of the rational and orderly. Where abnormality is not only accepted but unavoidable.

Depending on how it is defined, the Driftless Area embraces between sixteen thousand and twenty-four thousand square miles, mostly in southwestern Wisconsin but also in portions of adjacent Minnesota, Iowa, and Illinois. The Mississippi River runs through it, as do its feeder streams: the Saint Croix, Red Cedar, Cannon, Chippewa, Zumbro, Whitewater, Trempealeau, Black, La Crosse, Root, Pine, Bad Axe, Upper Iowa, Baraboo, Kickapoo, Wisconsin, Yellow, Turkey, Grant, Platte, Pecatonica, Sinsinawa, Galena, Maquoketa, Apple, and a thousand other smaller rivers and creeks, rivulets and cold springs. Geologists describe the characteristic pattern of the Driftless river drainages as *dendritic*—like the branching of a tree or the fingers on our hands, like the splayed-out interconnecting ends of our neurons.

The Driftless Area has other names. Geologists sometimes refer to it as the *Paleozoic Plateau*. Some call it the driftless *zone* or *region*. Around La Crosse, Wisconsin, its municipal heart, people speak about the "coulee region" (from the French Canadian *coulée*, from the French *couler*, meaning "to flow"). In other areas it goes by *Little Norway* or *Little Switzerland* (reflecting segments of its European settler demographic).

Here's the key thing to know about the Driftless: it defies the common image of the American Midwest. Because the landscape was never leveled by the glaciers of the Pleistocene, it is not pancake flat. You can't drive straight through it at eighty miles an hour on the way to Denver. It slows you down. It makes you turn.

The Driftless is an anomaly. Through the recurring episodes of Pleistocene glaciation—seventeen pulses of expansion and shrinkage over two and a half million years—ice hemmed in the Driftless on all sides at one time or another but left its interior ice-free. To the east, Lake Michigan's north-south basin served as a sluice, channeling one great lobe of glacial ice through its periodic advances and contractions. To the north, the ice sheet dove into the depths of Lake Superior's bowl, while the hard bedrock highlands just south of Superior limited ice flow into what is now Wisconsin.

The southern flanks of the ice sheets were relatively thin, and even modest variations in topography were enough to influence the shape of the glacier's edge. To the west, the great ice had an open field—the flatlands of the midcontinent—to ease its way south. As the ice sheets advanced and receded, over and over, they scraped clean the high-latitude and high-altitude landscapes of the continent but missed the Driftless. The most recent advance maxed out some twenty thousand years ago, melted back, and left behind its burden of boulders, gravels, sands, silts, and clays—the "glacial drift." But this odd exception, this large dent on the southern margin of the North American ice, remained unglaciated and, hence, *driftless*.

\* \* \*

A few miles from the Haugen farm, on the western edge of Coon Valley, a roadside historical marker commemorates the revolution: "Nation's First Watershed Project." The text explains that this valley served as "the nation's first large-scale demonstration of soil and water conservation." In its own way, this marker might well be placed alongside those that stand at Lexington and Concord, Seneca Falls, Fort Sumpter, Little Big Horn, Pullman, Selma, Stonewall. Those places became emblematic of dramatic changes in our nation's human relations. Coon Valley, in the heart of the Driftless Area, was, and is, symbolic of far-reaching changes in our human-nature relations.

The Driftless Area is not a "pristine" wilderness. Humans have played a transformative role in the region ever since Paleoamericans, drifting along the edge of the receding glacier, searching for favorable hunting and gathering opportunities, came upon the great gap in the ice wall. Although the debates among New World paleontologists continue, it appears that the newly arrived humans and their descendants were complicit in the extinction of the mastodon and mammoth, the dire wolf and short-faced bear, the giant beavers and ground sloth, ancient camels and horses, and other Pleistocene fauna. Over the next dozen millennia, a series of Archaic, Woodland, and Mississippian peoples made their home in the Driftless, hunting and fishing, growing gardens, running fire through the prairies. In the later stages of prehistory, they inscribed their own distinctive marks on the land: the Driftless Area was the epicenter of the effigy-mound-building cultures of the midcontinent, their varied earthworks dotting the landscape in profusion (and still do, even after widespread destruction of mounds over the last two centuries). Modern tribes of the Driftless landscape include the Ho-Chunk, Sauk and Fox, Santee Dakota, Kickapoo, and Ojibwe. European explorers and missionaries came into the Driftless starting in the 1600s, to be followed by transient trappers, miners, loggers, and in the mid-1800s, immigrant settlers. By the 1930s, three generations of farming the Driftless ridges, slopes, and valley floors had brought a measure of prosperity but also an accelerating rural crisis in the form of ruinous soil loss, flash flooding, degraded woodlands, and depleted wildlife (as so distressingly exemplified at Coon Valley).

The Driftless Area is, then, a long-peopled and much-used landscape. And as with the rest of the planet, more than four hundred parts per million of atmospheric carbon dioxide (including the 120 post–Industrial Revolution parts) now waft over the coulees. Still, the earth endures and reminds: however changed and however constantly changing the landscape, it is not and will never be a completely humanized one. On the steepest slopes with the thinnest soils and driest conditions, remnants of the pre-European vegetation—"goat prairies" and oak savannas—still hold fast onto the outcrops. The sandstone, limestone, and dolomite bedrock, poking out of the hilltops like impacted molars, ground us in the nonhuman and prehuman cycling of carbon and minerals among atmosphere, ocean, and earth. The region has its share of dams and ditches and dikes, but the dendritic network of branched waterways still utterly defines the region.

And it was the way of water that finally forced necessary changes in land use in the 1930s. In the face of destructive floods, gullied slopes, sloughing soils, and dissolving pastures, people in the Driftless had to *make a turn*. Of all the restorative measures that the region's landowners adopted, and the many that have been retained since, the most readily visible are the alternating contoured strips of crop and pasture, hayfield and woodland edge, that hug the Driftless hills. Retaining soils, recycling nutrients, interrupting the gravity-pull of water downhill, the contours are nowhere uniform; they are unique to each piece of land, expressing its Paleozoic past, its land-use history, and its contemporary land ownership. Each parcel tells a tale of a farmer willing, at some point, to counter convention—perhaps even a neighbor, a friend, a father—to change from plowing straight up and down the slope to following the lead of the land and turning with it. So basic, and so radical. Such a wild thing to do.

If the Driftless Area is not "pristine," nor thoroughly humanized, neither is it like the rest of the agro-industrial American Midwest. It is not wholly engineered to serve as a mere medium for corn and soybeans bound for the global market. It has not been made efficient to the point of diminishing returns. The goat prairies, woodlands, bottomland forests, riparian wetlands, rivers, streams, and springs keep the landscape diversified. Smaller-scale dairy and livestock operations, with actual grazing animals, remain relatively viable so that a large portion of the land is covered in permanent pasture. The corrugated topography does not lend itself to ever-expanding economies of scale. Even the big-box stores have a hard time squeezing into the narrow valleys. Whatever algorithms allowed Walmart to proliferate with surgical precision, conquer the flat Midwest, and redirect the flow of capital, they presumably had to be rejiggered in the Driftless.

Like all places, then, the Driftless Area landscape is a complex expression of natural features and processes that are always shaping, and being shaped by, human actions that began long ago and that continue up to this instant . . . including actions unforeseen even a few years ago. The near-surface sandstones so characteristic of the Driftless now make the region ground zero for the extraction and processing of "industrial sand," an essential ingredient in the hydraulic fracturing ("fracking") process. The modest economy of the region makes the prospect of quick frac-sand profits attractive to many landowners and local municipalities. The global economy—and the fossil fuel juggernaut that feeds it—leaves no place untouched. Here, it scrapes land bare in a way that seventeen onslaughts of glacial ice over two thousand millennia could not.

And so the human impressions on the land emerge and fade, accelerate and slow, intensify and wane. Since Joseph Haugen was a boy—since the Haugens and the other farming families along Coon Creek signed up for the watershed restoration project—the Driftless landscape around him has changed. In many ways, it's grown wilder. Its soils are healthier, more stable, more productive (agriculturally and ecologically). Its surface waters, slowed in their overland flow, clarified by infiltration, chilled by their passage underground, now support thriving populations of trout (and a thriving fishing economy). Its remnant prairies and savannas are treasured. In the last two decades, black bears have come into the region from the north in increasing numbers. Gray wolves have reestablished themselves along the northeast edge of the Driftless, with occasional dispersers crossing over and testing the levels of human tolerance. Phantomlike, cougars come and go amid the coulees, caught on trail-cams as they arrive from as far as the Black Hills, and head off stealthily to points east.

Even as the American Midwest was surveyed and settled, gridded and sodbusted, plowed and ditched, simplified and commodified, the Driftless Area in its midst took a different path. The patterns and methods of land exploitation that worked so smoothly in the flatlands—that assailed the native flora, fauna, and peoples, that disrupted the region's soils and waters, that imposed supposed efficiencies — met their match in the convolutions of the Driftless. Here, the main stream of culture had to self-correct. Here, that culture had to admit to itself that self-correction was in fact called for and that progress does not always entail going full bore, heedless, straight ahead.

Over the last decade, several "five-hundred-year" floods have come to portions of the Driftless. The Haugen farm was among those in the path of several epic rainstorms, intense downpours of the sort that are expected to become more common with accelerated climate change. Even the professional soil and water conservationists who most closely monitor these rain events were surprised and encouraged to see how well the watersheds responded. The conservation measures first adopted seven decades before did their job — performed, in fact, beyond their design specs. Here, where the nation's first watershed project was undertaken, we learn vital lessons for the uncertain future: as we ignore the particular qualities, needs, and opportunities of the land, we put ourselves at risk; as we work with the wild, the land grows more resilient; and as the land grows more resilient, so do our communities.

\* \* \*

Throughout its history, the Driftless Area has regularly attracted renegades, refugees, resisters, and adventurers. Ho-Chunk who were removed time and again from their homeland but whose love of the land kept them returning. Fur trappers from France. Lead miners from Cornwall. Homesteaders from out east. Quakers who came in shortly after Wisconsin became a state in 1848. German "Forty-eighters" and Scandinavian farmers. Escaped and freed slaves who, before and after the Civil War, built their own community, Pleasant Ridge, in Grant County, Wisconsin. Black Hawk and Frank Lloyd Wright and Aldo Leopold. Since the 1960s, the Amish have come into the Driftless, drawn by its rural character and relatively affordable farmland. The Driftless remains a tolerant home to the unconventional: independent farmers, seed savers, organic growers, aging hippies, young agrarians, outsider artists, Wiccan worshippers, unpredictable voters, river rats, trout bums. Anomalies all.

Another visit, another farm.

Just a few miles away from the Haugen farmstead, and just a couple years before, I am at work with a film crew documenting a bit of Coon Valley's conservation history. My filmmaker friends Steve and Dave — from California and Colorado, respectively — are new to the Driftless. We ask the indispensable Jon Lee if he can help us locate an Amish farm where we might be able to film. We do not want to impose, especially on a mellow midday in early October—prime harvest time. But soon we find ourselves on another ridgetop farm. The farmer is hitching his team of brown-and-white Paint/Percheron horses, preparing to bring in hay and oats. He agrees to let us film but requests that we not record any closeup images. He also asks us, by way of barter, to help him pitch a load of straw bales.

After finishing the chore, we wait for the farmer to harness his team and come around the fields. We stand gazing across the valley, where the local Amish schoolhouse sits on an adjacent ridge. Recess has just been called. Fifteen boys and girls, dressed in brown and blue, bonnets and suspenders, emerge and commence playing baseball. We watch with fascination. A right-handed pull hitter has it made: one line drive into the clover, in the steeply pitched Driftless left field, and the ball will roll on until it reaches the Mississippi River. We listen to the click of bat on ball, the laughter of the children, the rustling of the leaves in the autumn breeze, the whinnying of the horses behind us.

During a pause in the action, Steve offers color commentary. Then he says, balanced perfectly between joking and seriousness, "I have never felt so American in all my life!"

A bucolic moment, caught on memory, framed by the billowing hills and odd angles of the Driftless and by the unsettling tensions and restless discontents of our times. But even the Amish—*especially* the Amish—are nowhere near as simple as they appear to be. It is not a simple life that can defy pressures to conform, or simple convictions that can maintain modesty. It is not simple routine that allows a man to milk cows for forty-seven uninterrupted years, or simple warmth that allows an octogenarian to smile when strange students come knocking, or simple need that causes farmers and tractors to turn with the contour. It is not a simple notion of the wild, or the human, that brings us around. We try to impose our will, yet we are shaped fundamentally by the wild, the spontaneous, the nonhuman, by forces that are greater than us, by realities that are older than us, by futures that draw us out. We are always finding ourselves on the edge of anomalies. And anomalies—with proper care and cultivation, exploration and contemplation, coordination and action—can seed revolutions.

## *References*

Anderson, Renae. "Coon Valley Days: A Short History of the Coon Creek Watershed." *Wisconsin Academy Review* 48, no. 2 (2002): 42–48. http://www.nrcs.usda.gov/Internet/FSE_DOCUMENTS/nrcs142p2_020471.pdf.

Hawkins, Arthur S. "Return to Coon Valley." In *The Farm as Natural Habitat: Reconnecting Food Systems with Ecosystems*, edited by Dana L. Jackson and Laura L. Jackson, 57–70. Washington, DC: Island Press, 2002.

Heasley, Lynn. *A Thousand Pieces of Paradise: Landscape and Property in the Kickapoo Valley*. Madison: University of Wisconsin Press, 2005.

Hubbuch, Chris. "Ernest Haugen: Farmer, Conservationist Was 'Icon of Community.'" *La Crosse Tribune*, November 12, 2011. http://lacrossetribune.com/news/local/ernest-haugen-farmer-conservationist-was-icon-of-community/article_e4672750-0ce3-11e1-956f-001cc4c03286.html.

Johnson, Hildegard Binder. *Order Upon the Land: The U.S. Rectangular Land Survey and the Upper Mississippi Country*. New York: Oxford University Press, 1976.

Knox, J. C. "Human Impacts on Wisconsin Stream Channels." *Annals of the Association of American Geographers* 67, no. 3 (1977): 323–42.

Leopold, Aldo. "Coon Valley: An Adventure in Cooperative Conservation." *American Forests* 5 (1935): 205–8.

———. "Wilderness as a Form of Land Use." *Journal of Land and Public Utility Economics* 1, no. 4 (1925): 398–404.

Lyons, Stephen J. *Going Driftless: Life Lessons from the Heartland for Unraveling Times*. Lanham, MD: Rowman and Littlefield, 2015.

Meine, Curt, and Keefe Keeley. *A Driftless Reader*. Madison: University of Wisconsin Press, forthcoming.

Trimble, Stanley W. *Historical Agriculture and Soil Erosion in the Upper Mississippi Valley Hill Country*. Boca Raton, FL: CRC Press, 2012.

# 5

## ORDER VERSUS WILDNESS

*Joel Salatin*

A member of the Virginia Monarch Butterfly Society called me: "Do you know where we can plant a pallet of milkweed seed?"

I didn't even know Virginia had such an organization. Beyond that, I wondered where in the world they procured a pallet of milkweed seed. As I talked with the lady on the phone, I suppressed my laughter, realizing that a couple of hours before I had had a totally frustrating in-the-field meeting with the landlords of one of the farms we rented. The landlords were more than a little dismayed at the weeds we had created with our mob grazing management. In September, and right when the monarch butterfly larvae needed them, those weeds included a healthy contingent of seed-pod-bursting milkweeds. The monarchs were euphoric. The landlords weren't.

This ninety-acre pasture farm had been continuously grazed for years before we rented it. The sparse grass never exceeded a couple of inches in height; clover was virtually nonexistent; thistles dominated the plant profile. In three years, by using mob grazing and aggressive hand tools, we vanquished the thistles, but a plethora of edible and often delectable weeds (like milkweed) thrived.

Indeed, that afternoon at our pasture-based summit, Daniel (my son) and I exulted in the biomass volume we'd stimulated. Fall panicum, milkweed, redtop, clover, and some goldenrod offered color and variety to the orchard grass and dominant fescue sward. The landlords, however, did not share our euphoria. As we stood in armpit-high biomass, arguably more than had been on the farm for decades, all the landlords could utter was a contemptible and emphatic "Look at all these weeds."

I was incredulous. Outdoor and wildlife lovers, the landlords could not make the connection between this diversified, voluminous biomass and the overall health of their pasture farm. We could scarcely walk through the biomass jungle, replete with spiders, field mice, and a host of creepy-crawly insects. They wanted a mono-species golf-course look. We wanted full expression of as much diversity as possible. These two objectives could not have been more incompatible. Unable to come to a consensus on landscape objectives, Daniel and I walked away disappointed, terminating the lease. Life is too short to spend every day arguing.

Later that evening, the monarch butterfly lover and milkweed propagator called. How ironic. I assured her that she didn't need to plant milkweeds. They love Shenandoah Valley pastures in the fall. I explained that all you had to do was create a pastoral mosaic by moving the herbivores every day to a new paddock. The recently grazed, waiting-to-be-grazed, and recovering-from-grazing paddocks encourage multispeciation and ongoing plant expression throughout the year. In other words, rather than having all the acreage at the same height or physiological point, different paddocks in different stages of growth exhibit a far more quilt-like pattern.

I explained that if the monarch society would spend its efforts encouraging mob grazing, the butterflies would have plenty of wild-grown milkweeds to enjoy. Since cows and sheep like to eat milkweeds, in a continuously grazed setting, the milkweeds never mature to seed stage — they are grazed extremely young as soon as they shoot up above the close-cropped sward.

Although I couldn't see her, I could tell that my excitement regarding mob grazing to make healthier butterflies was too far a stretch for her to comprehend. After all, she had a pallet of seed sitting in her garage that needed to be planted. That was the immediate need of the hour, thank you very much. Don't bother with trifles. I can only imagine her discouragement when talking to fellow butterfly lovers that she called the most outspoken ecological farmer in the state and got fed a bunch of gibberish about cows, weeds, and butterflies. What rubbish.

These two conversations on the same day illustrate the tension between order and wildness. I think our culture suffers from the perception that ordered farming, or ordered landscapes, must inherently militate against wildness. Indeed, Henry David Thoreau, as much as I appreciate most of his observations, perpetuated this idea of wildness as separate from human intervention. Perhaps he did not realize at that time, like we do today, the extent of landscape manipulation his area had been subjected to for millennia before European arrival.

Indeed, the very soil fertility the colonists enjoyed had not been built with dark, foreboding forests. These fertile soils developed under silvopastures meticulously maintained by migrating herbivores, predation (both two-legged and four-legged varieties), and strategic fires. The deep, dark, brooding eastern forests of Ichabod Crane mystique were a European invention, not Native American, as we now know. To be fair to Thoreau, in his day the tools to replicate wildness in domestic farming did not exist: tools like electric fencing, plastic water pipes, nursery shade cloths, and Tinkertoy-dimension lumber for lightweight, portable shelters.

Can we have wildness without migratory animals and free-roaming fires? Here in Virginia, carving out a state park could not be farther from true wildness. If such a park could reinstitute marauding migratory herds and bird flocks big enough and thick enough to blot out the sun for three days (documented in Audubon's journals), along with routine fires, perhaps a semblance of nature's wildness could occur. But barring those parts of the recipe, the designated park is an entirely different concoction. Many nature lovers ignorantly think that a designated spot of human abandonment replicates yesteryear's landscape. Such perceptions are naïve.

The idea, perpetuated by Thoreau, that farming order and wildness were mutually exclusive and therefore required segregated and designated areas allows landscape managers to be lazy about wildness. Perhaps lazy is too strong a word. But I find it disconcerting that too many farmers, arguably the largest landscape managers, retreat to this segregated mentality just like the radical natural park folks. I'd like to see more creativity, more visceral expressions, of commercial farming order not only coexisting with wild systems but actually enhancing them. Can this be done?

Even in the pages of *ACRES USA* we see pictures of clean tillage with mono-crops. Using compost and foliars, eco-farmers outperform their chemical-based counterparts. That's good and I'm glad. But can we do better?

A couple of years ago, I visited Colin Seis's farm in Australia. Standing in a two-hundred-acre field of oats ready to combine, imagine my surprise at seeing a healthy foot-high stand of diversified prairie-type forages growing under the oats. Inventor of what he calls pasture cropping, Seis has quickly attained worldwide attention for using livestock as an herbicide to weaken an existing perennial pasture enough for a crop of annuals to germinate and get ahead of the pasture.

The native pasture remains suppressed under the thickening annual crop canopy and finally begins to grow again once the crop begins drying down. By the time the combine comes through and harvests the grain, the pasture is a foot high and growing aggressively. There are two keys here. The first is using controlled grazing (duplicating migratory herds) strategically to temporarily weaken the existing sod. Second, planting does not require clean tillage. The result is arguably less ordered than a successful crop farmer may enjoy, but it's far more wild-oriented.

Field mice still have a place to live. Earthworm burrows remain unmolested. Soil is neither stirred nor inverted. Neither is it herbicided. The point is that the wild ecosystem coexists with the ordered crop program.

Last year, encouraged by what I'd seen at Colin's, on one of our rental properties we planted about thirty acres of corn, winter peas, and milo in the late spring into an existing sod that we double grazed to knock back the forage. We had tremendous germination of the corn and peas. The milo struggled, perhaps because the seed was less aggressive. At any rate, the pasture remained robust but stifled underneath as the canopy closed in. Where we had good fertility, the annuals created a veritable jungle. The peas trellised up the corn, and when we turned the cows in to graze it, they ate every stalk and leaf and even pulled up some of the corn by the roots.

Some stray weeds and summer annual grasses grew up as well, giving a look of random disorder in this sea of biomass. When the cows grazed it off and left a sheet of manure and urine, the perennial grasses responded in kind. This system used no chemicals and no plow. I think we'll do this again.

Foresters in the Appalachians express growing concern about oak health. The most pessimistic experts say it succinctly: "Our oak forests are dying." Most silviculturalists agree that the problem is a lack of periodic disturbance. This used to happen with fire and large migratory fauna. Today we have neither. Can we duplicate it?

For several years now on some one hundred acres of forest, we've been running pigs. Using electric fences as a control, we can encircle a five-acre area and put in fifty hogs for a couple of weeks. This disturbance encourages new species of plants and reduces pest attacks on the trees because the pigs eat worms and bugs around the root collar. As the pigs open up areas and encourage forage under the trees where prior to their arrival only a static leaf or pine needle mulch covered the ground, wildlife proliferates.

Attracted to the more open, diversified landscape, wildlife thrives along with the trees.

Perhaps nothing attracts wildlife more than riparian areas. Water is not only the lifeblood of a farm; it's the lifeblood of wild areas. Imagine a few centuries ago when millions of beavers from New Mexico to New England made massive shallow ponds across the landscape. We don't have beavers like that, but we do have track loaders and other excavation equipment. In fact, we can build ponds more strategically on the landscape (higher) than beavers ever could.

Permaculture has done a wonderful job of encouraging water impoundments on the landscape. From waterfowl to thirsty deer, riparian areas attract wildlife while providing landscape hydration opportunities. Whether used for livestock water or irrigation, ponds create integrated wild-ordered models. You can indeed have your cake and eat it too.

I can't imagine why any farmer would put in pretty lawn fences before having functional ponds. We in eco-agriculture should lead the way in illustrating a wild-friendly agrarian model. Most environmentalists I encounter still actually believe that wildness must be sacrificed in order to have productive agriculture. They are duped by the industrial/mechanical agriculture mind-set that preaches intensification through mono-crops and factory animal houses to supposedly free up land for wildness (or wilderness, depending on the legislation of the day).

This is a tension that need not exist. My brother, Art, an airplane mechanic by profession, has built a thriving honeybee business at Polyface. Even though apiarists document all sorts of maladies today, his bees are doing quite well. He doesn't feed them sugar water, never moves the hives around, and leaves enough honey for them to overwinter. And one other thing: because our pastures are diversified and uneven in age, we always have something blooming.

If we as a culture actually farmed like wildness mattered, we would not be transporting pollinators from place to place. Moving bee yards across the country is symptomatic of single-species, single-aged production. Wildness depends on internal balance for security. In wild systems, nobody injects energy, plants, or animals into the system; it must be a viable community by its own synergy. What if we simply agreed that a food system that requires moving bee hives from place to place is an assault on nature? What would man-made farming order look like if it respected the order of the bees?

It would look like a mosaic of diversity in real time. It would be a cornucopia of uneven-aged plants offering pollinators blossoms throughout the season. It would be a place where animals and plants thrived together rather than in exclusive designated spots. It would be a dramatically interesting landscape rather than a monotonous one.

Submitting our perceived order to the order of the bees, I suggest, is at least a starting point in nesting our farms into wildness. Why must farming and wildness be mutually exclusive?

That is not to say that on our farm we don't shoot coyotes that kill chickens. Nor does it mean we're forgiving toward a she-coon who teaches her babies how to kill laying hens. Humans are part of the ecology too, and so we participate viscerally and actively, like we have since the beginning of time. We are part of the balance.

An urban area that bans deer hunting, for example, has no clue about how wildlife functions or how the predator-prey relationship works to maintain wildness balance. I find it disconcerting that folks exist who send money to an environmental organization to maintain wildness somewhere but at the same time hire a landscape service to spray their lawns with herbicides and chemical fertilizers to ensure an ordered domestic habitation. This is intellectual schizophrenia.

I see it among farmers as well. Even in my own decision making. It's almost as if we validate our human importance by reveling in the extremes of mechanical precision toward biology. Jerry Brunetti has done us all a favor by documenting, over and over, that brush and weeds often out-nutrition farmer-preferred plants. Those overgrown fences actually provide tonics and supplements, not to mention places for birds to nest.

Landscapes should fascinate the eye and the mind through varietal symbiosis rather than through the hubris of human accomplishment. We work in tandem with nature. We are colaborers, not tyrants or even masters. When Joel Arthur Barker introduced the world to the word *paradigm* a couple of decades ago, he noted that paradigms appearing to reach perfection are on the point of collapse. I can't help but think of this principle when I see the supposed panacea promise of genetically modified organisms—which arguably represent the current pinnacle of human-imposed order upon genetic wildness (randomness?)—breaking down with incriminating study after study.

We can all thank people like Jeffrey Smith for keeping us abreast of the latest GMO takedowns. That many humans actually think we can penetrate biologically alive DNA wildness with inanimate mechanical precision makes thinking people shudder. This epitomizes our fixation on order.

Nature's order certainly trumped man-made manipulation with the rise of bovine spongiform encephalopathy. Feeding dead cows to cows violated every natural pattern observable in the wild. Bison don't eat dead bison. Wildness presents the patterns, and we humans should leverage our creativity and mechanical inventiveness to caress these patterns. But override them we dare not.

In my own lifetime, I've seen an entirely new lexicon of warning signals. Thankfully, the development of indoor plumbing, electricity, and stainless steel brought the infectious disease symptoms of unhygienic living under control. Now we're fighting both a rising new tide of noninfectious diseases and a new high-pathogenic problem: *Salmonella, E. coli, Campylobacter, Lysteria*. From food allergies to porcine virulent diarrhea (now killing every fourth piggie born in the United States), nature's wildness reminds us that hubris-based order must always submit to a design that we don't necessarily understand.

Why is a pig the way a pig is? Why is a violet the way a violet is? Ultimately, it's a mystery. As eco-farmers, we embrace that mystery for magnificence beyond comprehension. Rather than conquering, we submit. We work with wildness, not against it. Indeed, we should be wildness allies, weaving into our farmscapes the mystery and patterns wildness exhibits. That is our high calling and sacred mission.

# 6

## BIOMIMICRY

## *Business from the Wild*

*Margo Farnsworth*

> No man is an island,
> Entire of itself.
> Each is a piece of the continent,
> A part of the main.
> —John Donne, "Meditation XVII"

John Donne writes of the inescapable connection among all humankind, and yet his poetry is eerily incomplete. What are humans without birdsong in the spring? How can we consider ourselves independent of the food, medicine, housing, oxygen, and spiritual sustenance that wild organisms and spaces provide? Our relationship to the wild is not one of separation; we are "braided" together (Claus 2014, 46). Even a worm does not stand alone. What hubris to treat the world as if humans can.

Biomimicry is a tool we can use in business to recreate connections and perhaps even reconcile with the wild to a degree. Biologists, designers, engineers, and other business professionals are reconnecting to the wild with this tool by emulating organisms and even ecosystems, using nature's forms, processes, and systems as muse, teacher, and template.

But to mimic nature, one must first observe organisms and the forces acting on them that create a *genius of place*, that recipe of strategies that allows life not just to exist but to prosper in specific conditions unique to any geographic location. These strategies of the wild can allow us to optimize material use; build with life-friendly chemistry; and adapt to varied levels of moisture, temperature, pressure, and even oxygen. This genius of place is not a human space alone but a larger shared blend of wildness and order that together composes ecosystems and is interwoven with our lives. This braid of wild and nonwild exists in a tenuous balance because, too often, we have not adequately honored the genius of the wild and are seldom even sure how much wildness is enough to ensure a flourishing world. The practice of biomimicry ventures into new solution spaces that recognize the value of the wild while igniting hope for ways of doing business that closely attend to the genius of place.

Here is the story of one company's journey toward plaiting its nonwild existence to the wild right outside the door—reconnecting it to the main.

\* \* \*

David Oakey, of Oakey Designs, was frustrated. Ray Anderson, CEO at Interface, Inc., the world's largest manufacturer of modular carpet tiles and his main client, had issued a new vision for the company regarding sustainability.

"If we're successful," Anderson stated to employees and contractors, "we'll spend the rest of our days harvesting yesteryear's carpets and other petrochemically derived products, recycling them into new materials, [and] converting sunlight into energy, with zero scrap going to the landfill and zero emissions into the ecosystem. And we'll be doing well . . . very well . . . by doing good. That's the vision."

David was prepping for a three-year contract with Interface when Ray decided the carpet company was going to become sustainable. A carpet company in an industry where waste and dependence on petrochemicals was the rule of business—and now, sustainable? What did that even mean?

"It took me probably a year and a half of struggling to understand sustainability in our context," mused Oakey. His business mind was besieged by this new world of sustainability, the nexus where planet and people gained equal standing with profit. Nature's cycles seemed incompatible with the industrial textiles world he had always known. That world was born not in the Industrial Revolution but thousands of years earlier, when Asian shepherds first sheared their flocks and covered their floors with wool. Their human interest in overcoming and utilizing elements of the wild for clothing and home use or as space for crops, grazing, and expansion of permanent communities led to an ever-increasing alteration of wild places in lieu of existing in concert with them. Oakey hadn't yet considered the possible relevance to business of wild species and systems that had been adapting and retooling themselves for millions of years.

David feared that Ray's vision of sustainability meant designing with natural materials, which he knew would be cost prohibitive. Moreover, David thought, "Natural fibers don't perform like synthetics and even if they did, the amount of, say, sheep, we would need in Georgia alone for a growing population and carpet demand is mind-boggling." Oakey just couldn't connect the move to environmental sustainability with the economic realities required for this billion-dollar company.

However, Ray was successful, brilliant, kind, holistic — and dogmatic. After coming away from a company-wide meeting, David knew he had to either be on board with Ray's ideas or leave the company and go somewhere else. Sustainability — in whatever form — was going to happen here.

Ray had grown up in rural Georgia, a place where fragmented wildlands persisted. He walked to ball practice surrounded by nature's cycles. Towering longleaf pines and Shumard oaks shading his footsteps hosted hawks with a predatory focus on gray squirrels below. The squirrels, with watchful eyes and tails maniacally twitching, sported mouthfuls of Shumard bounty, while beetles systematically decimated the bodies of fallen oaks, returning them to the earth. Cycles within cycles.

But Ray's younger days were more oriented toward success in football, school, and business than attending to the ecological cycles that make the world and the lives of all its harried, varied inhabitants possible. It wasn't until 1994, after reading Paul Hawken's description of how humans have disrupted natural cycles and Daniel Quinn's description of "leavers" and "takers," that he was moved toward blending conservation with business leadership. And now, he was asking Oakey to do the same.

While David pondered this new challenge, over at the Interface Innovation Department, John Bradford, a farm boy who later added mechanical engineering to his resume, was breathing the satisfied breath of a businessman who has suddenly been freed to do better. As the chief innovation officer, John could now mesh his experiences from growing up on a farm with what he had learned as a mechanical engineer making carpet—

shooting him forward onto the then mostly blank canvas of sustainability in business.

In the early days of sustainability at Interface, John noted, "We started to really dream about what it meant for the company in the product and process aspects and in the inspirational innovation side of the business." John reflected on the times: "We started to really teach [the team], you know? Everybody read the same books. Everybody studied how to apply different streams of sustainable thought into different parts of the business."

The current director of innovation, Bill Jones, said it began simply "with the idea of using less and wasting less." Bill was working in the factory as the ideas started taking hold: "We'd been working with polylactic acid and continuously having struggles over the performance, availability, price, flexibility, and tenacity of the yarn. Then, we introduced a product with high postindustrial recycled content and achieved a 40 percent reduction on all seven measurements used in Life Cycle Assessment." Bill realized, "You're doing this! This is great!"

But simply reducing use of materials is not how nature manages or how it heals after disruption. It also wasn't Ray's idea of success. To adhere to nature's principles—to become restorative by reconciling manufacturing practices with nature's wild guidance—the company had to think and work in natural cycles.

Then one evening after one of Ray's public talks, he was approached by a scientist with masses of sunburned, chocolate hair who told him about a new way of working with nature and its cycles—in fact, all its rules—and how to do well by doing good. She told him about a process that would soon become a touchstone for him and his company. She told him about biomimicry.

In using biomimicry, one must observe the pas de deux between wild organisms and the forces—of wind, precipitation, altitude, and so on—that dance with them. The steps are crucial, because to miss one is to fall out of sync and bungle the whole affair. So it is with nature. The Biomimicry Institute, a world leader in bio-inspired practices, bases solution building on working in concert with what they refer to as "Earth's Operating Conditions": sunlight, water, gravity, limits and boundaries, dynamic nonequilibrium, and cyclic processes. These define wildness at the most basic level.

Most of us understand the importance of the first three operating conditions, whether we optimize their presence in our daily work and world or not. The fourth, limits and boundaries, is rather poorly understood. Humans regularly, sometimes proudly and flagrantly, violate limits and boundaries with little respect for how such actions might affect other organisms, including humans, or the planet as a whole.

"Dynamic nonequilibrium" is a key concept and yet is also frequently misunderstood because it is mistakenly envisioned as a pendulum-like process. A more apt description would involve an orange rolling around in a bowl—or an orange, as the eminent ecologist C. S. Holling put it, that experiences increasing amounts of third-party pressure (whether from a biotic or abiotic source) until the orange is suddenly released to shoot forward, rolling higgledy-piggledy around said bowl (see Holling and Sanderson 1996).

By understanding and applying the final element of cycles and cyclic processes, humans—including those in industry—can achieve some of the most potentially groundbreaking advances in sustainability. Consider the seasons with their cycles of rest and restoration. Birth, growth, and death with the subsequent birth of the next shiny new generation of fungus, plant, or animal provide another portrait of life cycles. "Waste equals food" (McDonough and Braungart 2002, 92) has become a key concept for the cyclical, circular economy, and increasing numbers of neighboring businesses are now using each other's waste as fuel for their own companies. But to optimize, whether in making the world turn or making a business profit, all of these "Operating Conditions" of the wild must be not only considered but understood and embraced.

With these foundational conditions, we can peer into nature to see how each organism optimizes with strategies that either work or pass from the earth.

During the fateful meeting between Ray and the young scientist who attended his talk, she suggested Ray's people go beyond their laboratories and conference rooms and out into the wild to reconnect and learn from nature's operating systems—and adopt those strategies of the wild.

A short time later, David was tasked with driving to the airport late at night to pick up that scientist, Dayna Baumeister. Dr. Baumeister had been hired to run a biomimicry workshop for a mélange of employees who represented every sector of the company, from the manufacturing floor to the marketing department. Dayna and her colleague Janine Benyus (who famously defined modern biomimicry in *Biomimicry: Innovation Inspired by Nature*) took turns taking the design, production, marketing, and business office team members out into the semiwild forests surrounding Interface's west-central Georgia offices. On these excursions, they asked the key biomimicry question: *How would nature . . . ?* For Interface, the question revolved around how to better design a floor. Out in the woods, gazing at the carpet of leaves, these team members saw an answer... diversity.

"If you look at anything in nature—leaves, flowers, rocks—it's diversity," David observed. "There is nothing that is the same. Making things the same and uniform is a human enterprise. So the fundamental principle that I started thinking about was, how can we make carpet tiles diverse, so that each carpet tile will come out different in color and design?" His voice suddenly tensed as he continued. "This was the complete opposite of what we'd been designing all my life. We'd been designing with uniformity. Each product would have to conform to the same color, same texture, same everything. We would call it 'quality,' but we were going against nature's grain."

The result was waste—waste on the cutting room floor, waste when dye lots didn't match, and waste when customers had to purchase extra stock to replace damaged or stained tiles. To counter such waste, David gave the challenge to Sydney Daniel, and together they hit upon a solution: a nonlinear/nondirectional-patterned carpet design soon to be named Entropy.

Meanwhile, over at the innovation department, biomimicry was following a different beat through Bill: "When I read the biomimicry book, I learned something different from what other people got about the amazing things that nature does. In nature, you fill a hole nobody else can, and that's efficient. That works." Bill's role took him down the path of science and math that would yield a profitable product. David's role was to use biomimicry for design and meet Bill at the end of that path.

Then one day there was a mistake.

"To minimize the amount of labor and down time with David's three-toned yarn diversity idea, I had tufted a large sample on the carpet-making mechanism that lets us test new tufting ideas. But we failed to put the directional arrows on the back of the tile so we could tell which way the true direction was," said Bill with a rueful smile. "Well, we kept laying it out and we weren't sure if we had the right runs or different runs mixed together."

Because this would affect everything from the look of the carpet to packaging, and ultimately customer satisfaction, they kept turning the squares, flipping them from one place to another. No one could tell which was the "correct" pattern. The chance for creating the same tile pattern was now about one in eight thousand. The result was not only that the squares would be easier to manufacture; they would also have less off-quality and be easier to package.

They had successfully engaged in two of the three practices required in biomimicry by both connecting or reconnecting with nature and emulating it. By mimicking the diversity they found, they had successfully emulated the forest floor. That allowed them to start complying with the operating systems of the wild.

"This was a kind of shallow foray into the world of biomimicry—looking at it from an aesthetics standpoint," John acknowledged. "But the more we dug in, the more we started to realize that it's about cycles." The more they applied nature's overarching strategies, the more they created a randomness of pattern that customers praised as comforting.

Not trained as psychologists but in need of broader insights about human well-being, the men started asking audiences who attended their public talks to imagine the place on earth that made them feel most alive. John remarked, "About 95 percent of the time they would say it was outside where they feel most alive. We don't really know the exact reason, but one of the things we've pinpointed is that humans try to control everything. We want to control color. We expect repeating patterns. We want there to be this 'exactness' about our interior space, but it's so different from our valued outdoor space. And this exactness affects us at a psychological level. . . . When you're outside, everything is random. Your brain isn't working overtime to pick up patterns. You're not trying to find flaws—because you accept that there really are no flaws. There are just these cycles of life around you. The wild can make mistakes, but its whole remains flawless."

He admits they never really put their finger on the psychological reasons underlying these preferences, but when they started working with natural models, a rhythm of randomness emerged—a rhythm that worked for the engineer, the mathematician, the designer, and the satisfaction of the customer. Biomimicry worked at the plant and the business office, too. Outcomes of their first venture into biomimicry included doubling their existing division's sales between 2002 and 2007.

Reduced waste also emerged as a benefit—from the 4 percent of standard carpet tiles down to 1.5 percent for the new products. There was no longer a need to reject mismatched dye batches. Recycling and disposal costs decreased significantly, and customer installation costs related to waste plummeted by 70 percent. Greatest of all, by avoiding the cost and waste of back-stocking replacement floor tiles, they saved customers $110 for every hundred yards ordered—making this the company's most popular line.

Here enters the third element of biomimicry in addition to connecting to and emulating nature: examining the ethics of human intentions and actions with the understanding that humans are only one among the millions of species on earth. Humans rely on the plenty provided by the natural world. Yet within this plenty a chasm exists between what we have known and understood and what remains mysterious and sometimes dangerous. In modern Western nations, most humans remain uncomfortable with this perceived split and work to subdue and conquer the wild, making it more understandable and useable and less wild in the process. Over time, we began to view extractive and unwilding practices as an inalienable right to be utilized at almost any environmental cost.

As John reflected on how changes were able to transform the company from a purely extractive entity into one that focused on reconciliation, he recalled his early university training: "Every single class I had in engineering started with a 'Given' statement that went like this: Assume that raw materials are abundant and consistent. Now think about that. How human is that—to crazily believe we are somehow going to continue to extract oil from the earth and make consistent materials around which we build an entire system? What if that's *not* true? What if that's a bad assumption?"

These were the questions the team began to ask. Their business had been built on arm's length transactions, a process where suppliers exchanged raw materials for cash. They made carpet and exchanged carpet for customers' cash. Because the average customer used the company's product for seven years and then sent it to a landfill, the team was experiencing dissonance. They were aiming for sustainability through biomimicry but engaged in a process that at its roots was unsustainable in the long term—both environmentally and economically. The dissonance also had roots in a simple principle that came from knowledge they attained in seventh-grade science: the earth is a closed system in which matter can't be created or destroyed. So when they created waste or customers were throwing away old carpet, it was still there—in a matted, darkened pile, but still there.

John continued: "So the ideal would be that we could all realize we are a species in a closed system. Everything that goes through us and every other species is going into our closed system. Then we could figure out, like nature, how to grow together in that closed system."

Knowing the realities of nature's laws and changing the direction of the charging bull that is business, which has historically aimed for the highest profits in the shortest amount of time, have been a matter of patience and vision. "As a society, we've got this insatiable appetite for more stuff—and it's all disposable," John observed. "But what if I went to my customers and my colleagues in my supply chain and developed a carpet leasing program? You can give those raw petroleum-based materials back to us and we can build our company more around the labor-heavy business of recycling than the raw materials–heavy business of creating virgin products year after year. It doesn't matter if a guy cares or doesn't care about sustainability . . . this isn't a belief thing . . . this is a straight-up numbers thing."

Short-term needs as a CEO of a company beget the tendency to think short term. A tightrope made of cost reductions pulling on one end and increasing income pulling on the other keeps the rope taut and the CEO aloft. Such an atmosphere begets the language of maximization. Nature's language is optimization. It uses only what it needs to build life. Interface had begun learning the language of optimization. They had Ray's vision of reconciling with nature as a means to do well *and good* in business to guide them as a North Star.

"A lot of times leaders and managers want to eliminate all of the reasons that something will fail before they decide to get behind it. What this does is tell all the people in the project that you don't believe in them. And Ray," John said, leaning in and dropping his voice, "would look for the light at the end of the tunnel, and when he saw it, he'd put his arm around you and say—*I'm bettin' on you, sport.*"

"That gives you a focus like no other focus," John nodded. "If you think your boss is that guy who's so afraid to step forward and so reluctant to believe in the human spirit that you don't take on the burden of success, then it's not yours. And if it's not yours, then you're just working for the man. And if you're just working for the man, it's just a job. And if it's just a job, then every other job is just a job, and what investment do you have in it? Ray got that. He got it down to his core."

The team didn't know when they walked out into the woods that the operating conditions discovered there would inspire their most profitable, fastest-growing line of carpets. But with Ray as a leader and a team of people that respected others for their roles on that team, Interface made biomimicry a flagship tool for reconciling with the wild and the operating conditions that drive it.

The team at Interface continues to work with nature's genius—for example, by integrating algae on a carpet background (since algae grows best on a 3-D substrate). The patented product is intended to filter and clean water at a variety of landscapes scales, from the Chesapeake Bay to an Arkansas hog farm.

They're also offering a business platform to lease carpet to customers, breaking the extractive cycle, putting more people to work, and expanding the carpet leasing program to enrich communities across the country.

The essence of reconciling business with the wild or semiwild can be found in the ongoing work to respect the roles of nature, its cycles, and its needs, which include our own. Instead of treating ourselves as a species apart, a community that need not heed "Earth's Operating Conditions," we can learn, as did Ray and his team, to braid our lives into nature—reconnecting to the continent as part of the main.

## *Acknowledgments*

In addition to listed references, this chapter is based on interviews conducted in 2014 with Interface staff John Bradford, Bill Jones, and Erin Meezan, and David Oakey of Oakey Designs. I extend my heartiest thanks to them, without whom this chapter would not exist.

## *References*

Beynus, Janine. *Biomimicry: Innovation Inspired by Nature.* New York: Harper Perennial, 1997.

Biomimicry 3.8. "Life's Principles." http://biomimicry.net/about/biomimicry/biomimicry-designlens/lifes-principles.

Claus, Anja. "A Language to Embody Place: Dynamic, Braided, Wild." *Minding Nature* 7, no. 3 (2014): 45–46.

Hawken, Paul. *The Ecology of Commerce.* New York: Harper Collins, 1993.

Holling, C. S., and S. Sanderson. *Rights to Nature: Ecological, Economic, Cultural and Political Principles of Institutions for the Environment.* Washington, DC: Island Press, 1996.

McDonough, W., and M. Braungart. *Cradle to Cradle.* New York: North Point Press, 2002.

Quinn, Daniel. *Ishmael: An Adventure of the Mind and Spirit.* New York: Bantam Books, 1992.

# 7

## NOTES ON "UP AT THE BASIN"

*David J. Rothman*

> ... Integrity is wholeness, the greatest
>     beauty is
> Organic wholeness, the wholeness of life and things, the divine beauty of the universe.
>     Love that, not man
> Apart from that, or else you will share man's pitiful confusions, or drown in
>     despair when his days darken.

The closing lines of Robinson Jeffers's mid-1930s poem "The Answer" cast a long shadow, providing the memorable title of the 1965 Sierra Club volume *Not Man Apart*. Edited by David Brower and with a foreword by Loren Eiseley, that volume set photographs of the Big Sur coast by Edward Weston, Morley Baer, Ansel Adams, and many others to verses by Jeffers as part of a successful effort to preserve the central California coast from encroaching development.

Yet Jeffers's poem is often misread. Many see it as a misanthropic sermon, which seems reasonable for a lyric that rises to a crescendo of rejection, eventually announcing that "A severed hand / Is an ugly thing, and man disseevered from the earth and stars and his history . . . for contemplation or in fact . . . / Often appears atrociously ugly." But then, in the passage quoted as the epigraph above, comes that sudden, intense turn away from his own hatred and toward "integrity," "beauty," and "the wholeness of life and things" and that cunning line break "love that, not man / apart from that," which is elided in the Sierra Club book title. Just at the moment when Jeffers appears to be telling himself to love only "organic wholeness, the wholeness of life and things, the divine beauty of the universe," he flips the choice from one that appears to be between that wholeness and man to one between that wholeness and man *separated from that wholeness*. And that is his "answer": not to reject man, but rather to realize that the rejection of man is as much an error as the other ugly errors men make. Such a rejection would only reiterate exactly the same separation from "organic wholeness" that the poem rages against.

This answer is something Jeffers needs to remind himself of forcefully, again and again, as he quarrels with himself over our apparent ugliness. Jeffers carefully, passionately balances his sickened response to human delusion, violence, ugliness, and evil (his own terms in this poem) against the realization that even such depravity is part of a greater and beautiful wholeness. The poem evokes both despair and a monistic, transcendent response to despair rooted in a vision of organic, universal, self-governing wholeness: the wild. His answer is to love a greater wildness of which we are always already part.

Throughout his work, Jeffers's vision of the wild and our place within it transforms the tradition of nature poetry. "The Answer," like many of his poems, asserts a worldview hardly ever expressed that way before him: that "the wholeness of life and things," which is transcendentally wild, is something we must imagine we are part of even to make sense of our own lives: "Love that, not man / Apart from that, or else you will share man's pitiful confusions." In this vision—an unprecedented vision—Jeffers integrates everything we are, even our ugliness, into that larger wildness, "the divine beauty of the universe." For if there is any wildness at all, how could the entire universe, considered as a whole, ever be anything but wild? It is certainly not domesticated, and we are but an infinitesimal part of it. It is a sublime, post-Hubble, poetic extension of Thoreau's observation that "In Wildness is the preservation of the world." Notice Thoreau's capital "W," the proper naming, that identifies something presumably beyond what can be named but yet must be named to be imaginable. Bedrock and paradox.

In the small wooded hills near where I grew up in Massachusetts, people clambered around dreaming of winter, cut a few trees, strung up T-Bars, built a ramshackle lodge, and gave names to trails—none of which mattered to the wild hills, as far as I know. But it mattered to us, even gave meaning to big swaths of our lives. What had once been just a hillside at a bend in the road that collected snow like a funnel now assumed a contour in our imaginations. It became a new geography where the simple act of making a figure in the snow integrated our lives with the hills, the woods, the sky, the season, and each other. It became a special place: a wild little forest hill with a name—"Snow Basin," for example.

Many of those areas closed long ago, and the woods have grown in. But what of it? What happened there? Why did Snow Basin matter so much to us? And what was that place? Where did we stop and that relatively gentle Wildness begin? Did it ever stop, once we came to recognize it? Were we only in it, or did it come to exist in us as well? What, exactly, were we making there? Turns? Friends? Races? Beauty? Memory? Wildness? Where has such a place gone now that we can only remember it? How do we quarrel with ourselves about the passage of time? How does one remember learning to love the Wild, that strange place we discovered and imagined on the hill and within ourselves? Has what we created there returned to something more wild than it was then? Is that wild something we might never have otherwise understood unless we had made something in it, with it, and of it within ourselves? These and other questions inspire appropriate wonder.

### *Up at the Basin*

Stan and Ruth Brown found the place in the 1940s,
When little ski hills were sprouting up everywhere,
A snow-belt hillside tucked up against a corner of Route 9
Just before it turns and climbs a hill to East Windsor,
A rolling hillside in the middle of nowhere.
But Ruth had raced in the big time and Stan was her coach,
And they understood how a good hill can coax gravity out of winter,
How even a bump can have "sporty terrain,"
As they put it, and they bought the land, cut a bunch of trees,
Eventually put in three T-bars, and gave the trails names
Like Sugar Run and West Loop, cut meandering through the maples,
And College Highway to Rumpus Run, a big boulevard down the front,
And Ruthie's, a joke on big old Ajax but a beautiful trail
Which ran down the eastern side of the hill,
And Steep Schuss, a sweet elevator pocket drop off Ruthie's.
The lodge was barely heated, and the floor was gravel.
Tickets $5 and everyone parked on the highway.
No snowmaking, no night lights, nothing fancy, good sticky buns.

Ruthie's hair gone white, still working every day.
One of the best races in my entire memory, top to bottom,
Ruthie's, Midway Steep, Spring Slope, another Aspen joke,
The Stan Brown Memorial Slalom, not homologated, but who cared?
I could reset the bamboo on it, turn for turn, today—
Set up and take the high line on the left-footer top of Midway,
Or else you'll blow the next turn, be crawling on the flats—
That's the way Dick set it, a good set, following the terrain.
We all stood around looking at that turn and talking tough.

Nothing there now. The area closed long ago,
Stan and Ruth gone, the lodge burned,
Just the old runs slowly growing in. Who would know
What we did there, the beauty of joy, from ruins?
But there's Marty, floating monster 720s off a jump
As big as a house that we built with a snowcat behind the lodge,
Looking like he's going to land in the parking lot.
And Rick, who I could never quite catch,
Arcing GS turns down Highway on his 207s, quiet and intense.
And Thaddeus, pounding bumps, so high he can barely talk.
And Dick, our coach, setting a course, shakes his head,
Muttering something about "that kid" under his breath.
And Ken, you ran the Patrol and probably more than that
And let us sleep in a spare room weekends,
Where one night in the next room at a party
There was an old guy named Bo, drunk as a dog,
Playing a washtub bass while some woman on her hands and knees
Sucked his big toe and everyone is laughing,
Yelling "Suck Bo's toes!"—can't quite figure out why.
And there I am, it's ten below zero and I'm running down Crescent Street
Then Henshaw Ave. carrying all my gear clumsy,
Running to meet the bus for the first day of camp.
It's Christmas break and all I can think is
We're going up to the Basin and I'm going to make the team,
I am going to be transformed. I will be a ski racer.
And the fierce king of winter himself
Is rolling across a crystal blue sky,
Announcing our incipient beards and biceps,
Celebrating the brief tenure of everything.

# PART 2

*Working Wild*

# 8

## LISTENING TO THE FOREST

*Jeff Grignon and Robin Wall Kimmerer*

Silently passing overhead, satellites capture photos of an anomaly in the landscape of northern Wisconsin, a great, dark block like an island in a sea of farms and fields. So distinctive in its sharp boundaries and unique thermal profile, the block is used by mappers as a georeference point. Imaging sensors utilize the latest technology, but there's much that science alone cannot see. The original aerial observer, the eyes in the sky of *Kenew* (the eagle), knows this land well. Riding the heated thermals over cornfields, *Kenew* dips low when he reaches the ancient pines and rides the cool contours of the softly domed maple canopy of the Menominee Forest.

Swaths of pines, drifts of maple over the rolling terrain—it wasn't always this way. Neither satellite nor *Kenew* can see through time; for that we need the storytellers, the deep memory of the oral tradition.

It is told that the Maple Nation has stood side by side with the Pines from the very beginning. Together these Elder Tree Nations watched over the Menominee Nation. The Menominee bear witness to the great changes in the environment over thousands of years, with the story held in carefully remembered words. Long ago, the three nations watched through dancing snow as the towering waves of ice appeared from the North. Riding in upon the snow and ice were the ancient forests of Yew and Spruce. The Elder Maples had responsibility to their plant communities who could not tolerate the cold, and with great sadness they moved south, away from the ice. After generations of hardship endured by the Pine and the Menominee Nations, the story tells of a change in the wind that signaled the great melting. When the land was flush with water, Maple Nation returned. From that point forward, the three nations were one. Since time immemorial, the people and the forest were inseparable.

Standing on the granite-slabbed banks, Jeff sweeps his arm over the Wolf River, white foam churning at our feet. "The Menominee have a name for our ancestors, *kiash mamaqcetaw*; the ancient ones. Our Creation story takes place right here," he says. "We have no migration story, because this is where we come from. The Menominee are the original inhabitants of this place. The forest we ardently reside in once stretched across ten million acres. For centuries the Pine, Maple, and Menominee Nations have lived off this land, making sure we each took only what we needed."

In comparison to the heavily settled areas that surround its borders, the Menominee Forest might well be classified as wilderness. The satellite flying above sees the wild river running through and the forest canopy unbroken for miles. With few people and development limited to scattered clusters of homes, this land stands out at night because of its darkness against a background spattered with artificial light. *Kenew* is witness to magnificent stands of big old trees, to the presence of intact food chains where bears wander and the night is marked by the howl of wolves. The Menominee Forest may look like a "wilderness" to the satellite, but *Kenew* sees it differently: a carefully nurtured web of reciprocity between people and land. It's a home.

We leave Jeff's well-worn khaki-green truck at the junction of several grassed-over dirt roads—just tracks, really—between young stands of yellow birch with golden bark shining in the spring sunshine. I don't imagine that anyone but Jeff has walked here in years, except for the deer and the bear who have dotted the path with their sign. There's no need for chatter as we push through the knee-high brush—the birds are talk enough. Without preamble, at a sign only he can see, Jeff steps off the path and pushes aside some young hemlocks with branches sweeping the ground. There's a whole family: tall ones, short ones, exotic yellow blooms floating on sturdy stems, with pale floral ribbons trailing beneath, a patch of rare yellow lady's slippers. This is why we've come, to visit them, to pay our respects, like walking to visit an old relative living back in the woods, just to be together, to renew the bonds. Jeff meets my eyes and just smiles.

"The Elder Tree Nations have been taking care of us from the beginning," Jeff says. The paths of the ancient ones are still known here and walked by modern Menominee. He takes a deep breath, looks around, and exhales, "Everything we need is given to us." This well-watered landscape brought well-being to the Menominee in the gifts of sturgeon and trout; maple sugar; wild foods, from leeks to berries to *manomaeh* in the wild rice lakes; deer; and good soil for their garden beds. The land is still a pharmacy, a library, a grocery store. It is also a place of deep ancestral connection, to language, to spiritual life. It is a sacred landscape where people and forest are linked by mutual responsibility. Jeff says, "For thousands of years the Menominees gave back the only way we could—in the forms of ceremonies, prayers, and offerings." The Menominee live in a culture of gratitude.

Gazing up at the sky, he continues: "The Menominee people are also called Dreamers. Through dreams and prayers, we told of the coming of the wooden boats. Shoulder to shoulder, People, Pines, and Maples watched as an erosive surge of people migrated from the East in search of land. These newcomers viewed our three nations as nothing more than a resource to be used. The White Pines fell to the timber barons in great waves, and the lands of the Menominee were reduced from 9.5 million acres of the ancient ones to 235,000 acres." Here the Pine, Maple, and Menominee Nations took shelter together within the boundaries of the reservation, refugees from a cultural climate change.

*Kenew* flew over a clear-cut landscape and in the precious remaining island of Menominee forest saw the people struggling to make their way in a changed world. Jeff knows the history of adaptation on this land and reminds us that "in order to save what was left of our nations, the Menominee people had to realize that the forest had not only spiritual and cultural value but an economic value as well that would help us survive. The forest taught us how we could harvest the trees in a fashion so that it will always be here, using a concept that says, 'If we cut the mature trees starting in the East, by the time we get to the west side of the forest, it will be time to start over.' This set up the concept of sustained yield, which is a modern forestry practice of today."

What the satellite sees as wilderness is intensively managed to provide livelihoods for people and at the same time to create ecological integrity. Most of the world thinks it can't be done, but Menominee practices show otherwise. The tired rhetoric of ecology versus economy, thinking you must choose between sustainability or prosperity, gives way here.

Jeff is a forester for Menominee Tribal Enterprises, the forest management arm of the Menominee Nation. His truck is full of maps and tools, but the most valuable tools he carries are the way that he sees and the way that he listens. He is in constant conversation with both his "babies" and his elders, the new generation of trees and the old ones standing by to lead them. Jeff draws from scientific forestry as well as traditional ecological knowledge to do his work. Unconfined by dogmatic intellectual monoculture, Jeff uses what he calls linear and nonlinear perception, the head and the heart, to understand the forest. He is nurturing the next generation of marketable trees by using Menominee principles of forest ecology that are taught by the plants themselves, honoring all parts of the community, from goldenrod to bark beetles. Some foresters see only merchantable stems and clear away other species that are seen as competitors. What they may try to exclude, Jeff includes, recognizing that each part of the ecosystem can be medicine for the whole.

At various times, Jeff has taken me tromping off to pay respect to the last remaining stand of American elms, refugees from waves of disease, or to check up on the progress of a patch of hemlock where regeneration seems to be weak, or to take note of the cone crop of ancient pines. Intimate knowledge of the community is a prerequisite for the careful silviculture of Menominee forest management. Each parcel of land is inventoried and measured again and again, and the trees are known as individuals—a remarkable level of familiarity for an area that spans 235,000 acres.

The familial relationship between people and the forest here calls into question the notion of wild. I asked Jeff how this might be expressed in the Menominee language. He explained that the idea of wild, untamed land did not exist in their world but came along with the surge of newcomers, who pressed up against the Menominee homelands, misunderstanding their nature. The Algonquin languages, to which Menominee belongs, contain the pronoun *Pekuac*, meaning "growing on its own," which indicates the freedom of that being to live where it will rather than the state of its landscape. He says that for him, "When you can feel the aliveness of everything around using all the senses, you are experiencing wilderness."

There is a widely told story of an anthropologist who asked the Native peoples he was visiting, "Do you go into the woods alone?" This scientist was reportedly baffled by the people's inability to answer this simple question, until he was informed that it was impossible to be alone where one is surrounded by nonhuman relatives. "Alone" has no meaning in this forest, and perhaps "wild" fades away as well.

The standard image of a forester, with chain saw and hard hat, does not usually include a person with an eye for wildflowers. But management of the vast Menominee Forest is guided in part by the blossoms that carpet the forest in spring. The foresters listen to what the presence of Trillium has to say or the message conveyed by Goldthread. They know that the forest floor plants are highly responsive to conditions of soil and moisture and so use them as indicators of the stand's future, following the guidance of the flowers in devising management plans. They harvest timber in winter when the snow protects the forest floor and lift, rather than drag, logs from the woods to avoid disturbance of the respected plant community.

Back in 1923, Huron Smith published a scholarly treatise on "The Ethnobotany of the Menominee Indians" based on the deep plant knowledge of a group of Menominee families. Jeff has devoted himself to relocating every species of plant on that list—visiting them and renewing relations, letting them know that they have not been forgotten. In this time of ecological losses, we might anticipate the emptiness of that task and resign ourselves to the dwindling of that list. How many places in this country still have the same species from a list that was recorded almost a century ago? But Jeff has found every single one—and they have found him.

The intimacy and knowledge of what the forest needs to thrive are told in part by scientific data. There is a grid of CFI (Continuous Forest Inventory) sampling plots in the Menominee Forest that provide a long and continuing record of tree growth and mortality. This is extremely valuable for knowing the forest and planning for its future. All the foresters rely on this, but there is a deeper knowledge available, held by the forest itself. Listening to the forest is a precious cultural resource that needs as much tending and regeneration as the forest itself.

Jeff is a regeneration forester, one in an unbroken lineage of forest caretakers, charged with nurturing a forest he will never see. His life is full of seedlings of all kinds, as he is also dedicated to the maintenance and regeneration of culture. He is a cultural resources protection specialist, too, identifying archeological sites for the nation with his keen ability to see into the scope of time and listen to the land with the tools of eyes and mind and spirit. He knows that you have to carefully nurture culture and language as carefully as you nurture new trees.

Sustainability of the forest is inseparable from sustainability of the culture. Preservation of the endangered Menominee language is key to the integrity of plant knowledge. Plant names in the language encode a whole host of meanings and relationships that defy translation to botanical Latin. It is said that as longtime citizens of this forest, the plants are shown respect by being addressed by their names, spoken in the original language. Seeds and saplings of cultural sustainability are everywhere at Menominee—in the school, as kids tend the sap boil at sugar camp, learning from the Elder Maple and the Elder Menominee at the same time; in the language immersion daycare, which Jeff's wife opened; at the traditional garden plots; in the Elders Center, where carriers of traditional culture are interviewed by tribal college students with video cameras; and in the forest, when a human sits quietly to listen to what the trees have to say.

Jeff and I take students from the tribal college out to a grove of elders where the signature voice of this place, wind in the pines, speaks loudly. Science classes are usually organized around learning *about* the forest, but here our intent is to learn *from* the forest. Sitting on a mossy log, he tells them, "We look to the teachings of long ago. The teachings tell us we must mend our relationships with the individual plant communities. They have been waiting. All the answers to the problems of today are found within the plants and their communities. We must learn again how to listen." The forest is full of teachers, who show by their ways that we are responsible for one another. The Elder Maples are looking out for their community; neither one flourishes alone, only together. Listen to the forest and you hear that every taking must be balanced by giving. It is this reciprocity, the exchange of forest gifts and human gifts, that keeps the balance.

While carefully nurtured tree by tree, the forest is subject to the forces of upheaval as well. The landscape is marked by at least two recent tornado tracks, which felled trees in a massive swath, shearing off the canopy and tearing up roots.

The Menominee know about drastic destruction. In 1953, the federal government decided to "get out of the Indian business" and the tornado of Termination swept across the nation. With the stroke of the presidential pen, the Menominee ceased to exist as a tribe. Their ancestral homeland since time immemorial was declared to be a county of Wisconsin, and all the services guaranteed and negotiated in treaty in recompense for the loss of 90 percent of their lands were cut off. A tornado of destruction—poverty, land loss, depopulation, and cultural fragmentation—ensued. Faced with the piecemeal sale of their sacred homelands, a group of strong young people arose in defense of their heritage and future. After eighteen years in the courts, the nation was legally restored, yet the real work of cultural restoration continues.

At Menominee, resilience has been called for over and over. The forest's strength is rooted in the solidarity of people and place. Strong cultural identity feeds that resilience, like nitrogen in the soil. Regenerating the community after Termination is a reassertion of sovereignty and self-determination, a statement of endurance and resilience, as taught by the forest.

In the aftermath of the tornados, forest regeneration began right away. Where the forest was sheared off, the elder Tree Nations met their responsibilities, providing shade and shelter to the remnants of the forest floor of Maple Nation. At the same time, the buried seed bank, newly exposed to the sun, released its long-dormant propagules to create a thick, fast-growing stand of fire cherry, aspen, and birch. Some foresters see these as "weed trees," but Jeff understands that they are essential to the healing of the land. These species add certain chemicals to the soil through their growth and their leaf litter that act as medicines for the healing of the forest.

Our forest ecology students push through the thicket of young trees, exclaiming over the seeming chaos of species that they find. There are members of the Elder Maple and Pine Nations, carrying the knowledge of what once was and what can be again. There are also members of the sun-loving Berry Nations and the Nation of Grasses, with fruits and flowers preparing the way for the future forest. A seed bank is a reservoir of possibility, an agreement with the future. It is the source of adaptation and resilience. It is made of stories and memory and dreams for the future. These young people are the seed bank of the nation.

In the aftermath of Termination, visionary leaders dreamed of the people the nation would need in the future: leaders with deep cultural roots and strong identity, cross-pollinating traditional ecological knowledge with the tools of science. And so they created the College of the Menominee Nation (CMN), preparing both the seeds and the fertile ground. Students who are strong in their own cultural values and also equipped with the tools of ecological science can be a source of resilience in an era of unprecedented environmental change.

Since the time of the ancient ones, the Menominee have always given gratitude, care, and respect back to the forest. *Kenew* has never failed in carrying these prayers skyward. As the times have changed, the forest that has always helped the people calls for their help in return. One student twists his hat in his hands and says, "The Forest takes care of us now, but what about climate change, won't that change everything?" Jeff meets his question with a grave nod: "Today the three nations stand shoulder to shoulder as we face a warming wind from the South. The temperature increase to Grandmother Earth poses one of the greatest dangers so far to our forest. In order to protect our nations, we have begun to use science guided by traditional knowledge. We are not conceding the loss of the Maple Nation; we are listening to the forest to find a way for them to stay in their ancestral homelands."

Jeff's battered old truck is pulled off the gravel road by a big clearing, a regeneration plot. He tends many such openings in the forest, nurturing the seedlings. Pointing out a beautiful community of young pines, he explains to the students that "the openings encourage a healthy turnover in age within the forest. Diversity of plant life is allowed to fill the gaps among the seedlings, plants being communal species. This diversity is dictated by the land and its needs. It will change over time as certain seedlings take on the responsibilities of an Elder. We encourage the reuniting of the Pine and Maple Nations within the openings. Each nation protects the other from insects and disease and in turn the Menominee Nation."

In other clearings, Jeff is planting temperature- and drought-tolerant oak species and their associated shrub and grass species—the Elder Tree community of the southerly forests—inviting them and their families here so that a century from now, when everything is different, these forest patches might be guided to a new kind of flourishing under the leadership of other trees. The forest may change, the climate may change, but the relationship endures.

From the coming of the ice to the coming of the timber barons, the Forest and the People have been resilient in the face of dramatic changes. Even after everything, they are both still here, still participating in an ongoing exchange of reciprocal gifts in which the forest sustains the people and the people sustain the forest.

Shovel in hand, Jeff gives each student seedlings to plant. They listen to his instructions and his closing words: "We as one people have to stop thinking of the environment as our resource to exploit and not having to give back in some form. In Menominee belief, we were the last in the four orders of creation: rock, plant, animal, and then human. We remember the words of our ancestors, *Kanawaenemaew kotapiskocekan ayom maeqtekuahkikiw, ahpihciwaeqtaw*—the strength of the environment is in its balance, and we are the balance."

# 9

## THE WORKING WILDERNESS

*Courtney White*

The only progress that counts is that on the actual landscape of the back forty.
—Aldo Leopold, "The Ecological Conscience" (1947, 338)

U BAR RANCH
*Silver City, New Mexico*

During a conservation tour of the well-managed U Bar Ranch near Silver City, New Mexico, I was asked to say a few words about a map a friend had recently given to me.

We were taking a break in the shade of a large piñon tree, and I rose a bit reluctantly (the day being hot and the shade being deep) to explain that the map was commissioned by an alliance of ranchers concerned about the creep of urban sprawl into the five-hundred-thousand-acre Altar Valley, located southwest of Tucson, Arizona. What was different about this map, I told them, was what it measured: indicators of rangeland health, such as grass cover (positive) and bare soil (negative), and what they might tell us about livestock management in arid environments.

What was important about the map, I continued, was what it said about a large watershed. Drawn up in multiple colors, the map expressed the intersection of three variables: soil stability, biotic integrity, and hydrological function—soil, grass, and water, in other words. The map displayed three conditions for each variable—"Stable," "At Risk," and "Unstable"—with a color representing a particular intersection of conditions. Deep red designated an unstable, or unhealthy, condition for soil, grass (vegetation), and water, for example, while deep green represented stability in all three. Other colors represented conditions between these extremes.

In the middle of the map was a privately owned ranch called the Palo Alto. Visiting it recently, I told them, I had been shocked by its condition. It had been overgrazed by cattle to the point of being nearly "cowburnt," to use author Ed Abbey's famous phrase. As one might expect, the Palo Alto's color on the map was blood red, and there was plenty of it.

I paused briefly—now came the controversial part. This big splotch of blood red continued well below the southern boundary of the Palo Alto, I said. However, this was not a ranch, but part of the Buenos Aires National Wildlife Refuge, a large chunk of protected land that had been cattle-free for nearly sixteen years.

That was as far as I got. Taking offense at the suggestion that the refuge might be ecologically unfit, a young woman from Tucson cut me off. She knew the refuge, she explained, having worked hard as a volunteer with an environmental organization to help "heal" it from decades of abuse by cows.

The map did not blame anyone for current conditions, I responded; nor did it offer opinions on any particular remedy. All it did was ask a simple question: Is the land functioning properly at the fundamental level of soil, grass, and water? For a portion of the Buenos Aires National Wildlife Refuge, the answer was no. For portions of the adjacent privately owned ranches, which were deep green on the map, the answer was yes.

Why was that a problem?

I knew why. I strayed too closely to a core belief of my fellow conservationists—that protected areas, such as national parks, wilderness areas, and wildlife refuges, must always be rated, by definition, as being in better ecological condition than adjacent "working" landscapes.

Yet the Altar Valley map challenged this paradigm at a basic level, and when the tour commenced again on a ranch that would undoubtedly encompass more deep greens than deep reds on a similar map, I saw in the reaction of the young activist a reason to rethink the conservation movement in the American West.

From the ground up.

## CS RANCH
*Cimarron, New Mexico*

My decision received a boost a few weeks later while sitting around a campfire after a tour of the beautiful one-hundred-thousand-acre CS Ranch located in northeastern New Mexico. Staring into the flames, I found myself thinking about ethics. I believed at the time, as do many conservationists, that the chore of ending overgrazing by cattle in the West was a matter of getting ranchers to adopt an ecological ethic along the lines that Aldo Leopold suggested in his famous essay "The Land Ethic," where he argued that humans had a moral obligation to be good stewards of nature.

The question, it seemed to me, was how to accomplish this lofty goal.

I decided to ask Julia Davis-Stafford, our host, for advice. Years earlier, Julia and her sister Kim talked their family into switching to holistic management of the land, a decision that over time caused the ranch to flourish economically and ecologically. In fact, the idea for my query came earlier that day, when I couldn't decide which was more impressive: the sight of a new beaver dam on the ranch or Julia's strong support for its presence.

The Davis family, it seemed to me, had embraced Leopold's land ethic big time. So, over the crackle of the campfire, I asked Julia, "How do we get other ranchers to change their ethics too?"

Her answer altered everything I had been thinking up until that moment.

"We didn't change our ethics," she replied. "We're the same people we were fifteen years ago. What changed was our knowledge. We went back to school, in a sense, and we came back to the ranch with new ideas."

Knowledge *and* ethics, neither without the other, I suddenly saw, are the key to good land stewardship. Her point confirmed what I had observed during visits to livestock operations across the region: many ranchers *do* have an environmental ethic, as they have claimed for so long. Often their ethic is a powerful one. But it has to be matched with *new* knowledge—especially ecological knowledge—so that an operation can adjust to meet changing conditions, both on the ground and in the arena of public opinion. Of course, a willingness on the part of a rancher to "go back to school" is a prerequisite to gaining new insights. Tradition, however, seemed to have a lock on many ranchers.

The same thing is true of many conservationists. In the years since I cofounded the Quivira Coalition, I came to the conclusion that it had been a long time since any of us had been back to school ourselves. Tradition was just as much an obstacle in the environmental community as it was in agriculture. It wasn't just the persistence of various degrees of bovine bigotry among activists, despite examples of healthy, grazed landscape like the U Bar, either. It was more a stubbornness about the relation between humans and nature—that they should be kept as far apart as possible—expressed in the long-standing dualism of environmentalism that said recreation and play in nature were preferable to work and use.

If conservationists went back to school, as the Davis family did, what could we learn? Aldo Leopold had a suggestion that can help us today: study the fundamental principle of *land health*, which he described as "the capacity of the land for self-renewal," with conservation being "our effort to understand and preserve this capacity" (1949, 221).

By studying the elements of land health, especially as they change over time, conservationists could learn that grazing is a natural process. The consumption of grass by ungulates in North America has been going on for millions of years—not by cattle, of course, but by bison, elk, and deer (and grasshoppers, rabbits, and even ants)—resulting in a complex relationship between grass and grazer that is ecologically self-renewing. We could learn that a re-creation of this relationship with domesticated cattle lies at the heart of the new ranching movement, which is why many progressive ranchers think of themselves as "grass farmers" instead of beef producers.

We could also learn that many landscapes need periodic pulses of energy, in the form of natural disturbance—such as fires and floods (but not the catastrophic kind)—to keep things ecologically vibrant. Many conservationists know that low-intensity fires are a beneficial form of disturbance in ecosystems because they reduce tree density, burn up old grass, and aid nutrient cycling in the soil. But many of us don't know that small flood events can be positive agents of change too, as can drought, windstorms, and even insect infestation. Or that animal impact caused by grazers, including cattle, can be a beneficial form of disturbance.

We could further learn, as the Davis family did, that the key to healthy disturbance with cattle is to control the timing, intensity, and frequency of their impact on the land. The CS, and other progressive ranches, bunch their cattle together and keep them on the move, rotating the animals frequently through numerous pastures. Ideally, under this system, no single piece of ground is grazed by cattle more than once a year, thus ensuring plenty of time for the plants to recover. The keys are regulating where cattle go, which can be done with fencing or a herder, and timing their movement, in which the herd's moves are carefully planned and monitored. In fact, as many ranchers have learned, overgrazing is more a function of timing than it is of numbers of cattle. For example, imagine the impact 365 cows would have in one day of grazing in one small pasture versus what one cow would do in 365 days of grazing in the same pasture. Which is more likely to be overgrazed? *Hint:* Have you ever seen what a backyard lot looks like after a single horse has grazed it for a whole year?

We could also learn, as I did, that much of the damage we see today on the land is historical—a legacy of the "boom years" of cattle grazing in the West. Between 1880 and 1920, millions of hungry animals roamed uncontrolled across the range, and the overgrazing they caused was so extensive, and so alarming, that by 1910 the US government was already setting up programs to slow and to heal the damage. Today, cattle numbers are down, way down, from historic highs—a fact not commonly voiced in the heat of the cattle debate.

A willingness to adopt new knowledge allowed the Davis family to maintain their ethic yet stay in business. Not only did it improve their bottom line; it also helped them meet evolving values in society, such as a rising concern among the public about overgrazing. Rather than fight change, they had switched.

As the embers of the campfire burned softly into the night, I wondered if the conservation movement could do the same.

## KAIBAB NATIONAL FOREST
*Flagstaff, Arizona*

A friend of mine likes to tell a story about the professor of environmental studies he knows who took a group of students for a walk in the woods near Flagstaff, Arizona. Stopping in a meadow, the professor pointed at the ground and asked, not so rhetorically, "Can anyone tell me if this land is healthy or not?" After a few moments of awkward silence, one student finally spoke up and said, "Tell us first if it's grazed by cows or not." In a similar vein, a Santa Fe lawyer told me that a monitoring workshop at the boundary between a working ranch and a wildlife refuge south of Albuquerque had completely rearranged his thinking. "I've done a lot of hiking and thought I knew what land health was," he said, "but when we did those transects on the ground on both sides of the fence, I saw that my ideas were all wrong."

These two instances illustrate a recurring theme in my experience as a conservationist. To paraphrase a famous quote by a Supreme Court justice, members of environmental organizations can't define what healthy land is, but they know it when they see it.

The principle problem is that we are "land illiterate." When it comes to "reading" a landscape, we might as well be studying a foreign language. Many of us who spend time on the land don't know our perennials from our annuals, what the signs of poor water cycling are, what leads to a deeply eroded gully, or, simply by looking, whether a meadow is healthy or not.

For a long time, this situation wasn't our fault. What all of us lacked—rancher, conservationist, range professional, curious onlooker—was a common language to describe the common ground below our feet. But that has changed.

In recent years, range ecologists have reached a consensus on a definition of health: the degree to which the integrity of the soil and ecological processes of rangeland ecosystems are sustained over time. These include water and nutrient cycling, energy flow, and the structure and dynamics of plant and animal communities. In other words, when scarce resources such as water and nutrients are captured and stored locally by healthy grass plants, for example, then ecological integrity can be maintained and sustained. Without them—if water runs off-site instead of percolating into the soil or grass plants die due to excessive erosion of the topsoil, for example—this integrity will likely be lost over time, perhaps quickly.

This is the language of soil, grass, and water.

Taking it to the next step, range ecologists echo Aldo Leopold's famous quote that "healthy land is the only permanently profitable land" (1946, 224). Producing commodities and satisfying values from a stretch of land on a sustained basis, they insist, depends on the renewability of internal ecological processes. In other words, before land can sustainably support a value, such as livestock grazing, hunting, recreation, or wildlife protection, it must be functioning well at a basic ecological level. Before we, as a society, can talk about designating critical habitat for endangered species, or increasing forage for cows, or expanding recreational use, we need to know the answer to a simple question: is the land healthy at the level of soil, grass, and water? If the answer is "no," then all our values for that land may be at risk.

Or as Kirk Gadzia, an educator, range expert, and coauthor of *Rangeland Health*, the pioneering 1994 book published by the National Academy of Sciences, likes to put it, "It all comes down to soil. If it's stable, there's hope for the future. But if it's moving, then all bets are off for the ecosystem." It is a sentiment Roger Bowe, an award-winning rancher from eastern New Mexico, echoes: "Bare soil is the rancher's number one enemy."

It should become the number one enemy of conservationists as well.

The publication of *Rangeland Health* was the touchstone for a new consensus on the meaning of land health within the scientific and range professional communities. It paved the way for the debut, in 2000, of a federal publication titled *Interpreting Indicators of Rangeland Health*, which provides a seventeen-point checklist for the qualitative assessment of upland health. A similar assessment has been made of stream health by a federal interagency group known as the National Riparian Team. The indicators of health include measures of the presence of rills, gullies, bare ground, pedestaling (grass plants left high and dry by water erosion), litter (dead grass, which retards the erosive impact of rain and water), soil compaction (which can prohibit water infiltration), plant diversity (generally a good thing), and invasive species (generally not)—the same indicators that formed the basis of the Altar Valley map that I described on the tour.

This was the message I tried to communicate to the young activist under the tree that hot summer day—that a rangeland health paradigm, employing standard indicators, allows all land to be evaluated equally and fairly. By adopting it, the conservation movement could begin to heed Aldo Leopold's advice that any activity that degrades an area's "land mechanism," as he called it, should be curtailed or changed, while any activity that maintains, restores, or expands it should be supported. It should not matter if that activity is ranching or recreation.

### CHACO CULTURE NATIONAL HISTORICAL PARK
*Southeast of Farmington, New Mexico*

In an attempt to understand the issues of land health better, I paid a visit to a famous fence-line contrast. This particular fence separated the Navajo Nation, and its cows, from Chaco Culture National Historical Park, a UNESCO World Heritage site and archaeological preserve located in the high desert of northwest New Mexico. Cattle-free for over fifty years, Chaco's ecological condition became a pedagogical issue some years ago when Allan Savory used the boundary to highlight the dangers in the park of too much rest from the effects of natural disturbance, including grazing and fire.

I wasn't a fan of fence-line contrasts myself, mostly because I dislike dichotomies represented by a fence: us/them, either/or, wild/unwild, grazed/ungrazed. The world is more complicated than that. I'd rather take fences down, or move beyond them. But fence-line contrasts have pedagogical value, especially for new students of range health—like me. I decided I wanted to see this contrast in particular, but I knew I needed help interpreting what I saw, so I asked Kirk Gadzia to come along.

Both of us were well aware of the park's history—that a century of overgrazing by livestock had badly degraded the land surrounding the famous ruins. We also understood that the era's typical response to this legacy of overuse was to protect the land from further degradation with the tools of federal ownership and a barbed-wire fence. That's how Chaco became a national park. At the time, it was a common and appropriate scenario that played out all across the West. But Kirk and I didn't go to Chaco to argue with history or to pick a fight with the National Park Service. We weren't there to offer solutions to any particular problem either. We simply wanted to take the pulse of the land on both sides of a fence.

We stopped along the road at the eastern boundary of the park (this was during the growing season). On the Chaco side, we saw a great deal of bare ground, as well as many forbs, shrubs, and other woody material, some of it dead. We saw few young plants, few perennial or bunch grasses, lots of wide spaces between plants, lots of oxidized plant matter (dead grass turning gray in the sunlight), and a great deal of poor plant vigor. We saw both undisturbed, capped soil (bad for seed germination) and abundant evidence of soil movement, including gullies and other signs of erosion. On the positive, we saw a greater diversity of plant species than on the Navajo side, more birds, more seed production, no sign of manure, and no sign of overgrazing.

On the Navajo side, we saw lots of plant cover and litter, lots of perennial grasses, tight spaces between plants, few woody species, a wide ageclass distribution among the plants, little evidence of oxidization, and lots of bunch grasses. We saw little evidence of soil movement, no gullies, and far fewer signs of erosion than on the Chaco side. On the other hand, we saw less species diversity, poor plant vigor, a great deal of compacted soil, fewer birds, less seed production, a great deal of manure, and numerous signs of overgrazing.

"So which side is healthier?" I asked Kirk.

"Neither one is healthy, really," he replied, "not from a watershed perspective anyway." He noted that the impact of livestock grazing on the Navajo side was heavy; plants were not being given enough time to recover before being bitten again (Kirk's definition of overgrazing). As a result, the plants lacked the vigor they would have exhibited in the presence of well-managed grazing.

However, Kirk thought the Chaco side was in greater danger, primarily because it exhibited major soil instability due to gullying, capped soil, and lack of plant litter. "The major contributing factor to this condition is the lack of tightly spaced perennial plants," he continued, "which exposes the soil to the erosive effects of wind and rain. When soil loss is increased, options for the future are reduced."

"But isn't Chaco supposed to be healthier because it's protected from grazing?"

"That's what people always seem to assume," said Kirk. "In my experience in arid environments around the world, total rest from grazing has predictable results. In the first few years, there is an intense response in the system as the pressure of overgrazing is lifted. Plant vigor, diversity, and abundance often return at once, and all appears to be functioning normally. Over the years, however, if the system does not receive periodic natural disturbance—by fire or grazing, for example—then the overall health of the land deteriorates. And that's what we are seeing on the Chaco side."

Then he added a caveat.

"Maybe land health isn't the issue here," he said. "It may be more about values. Is rest producing what the park wants? Ecologically, the answer is probably 'no.' But from a cultural perspective, the answer might be 'yes.' From the public perspective too. People may not want to see fire or grazing in their park."

But at what price, I wondered? Later in the day, we learned that the Park Service was so worried about the threat of erosion to Chaco's world-class ruins that they intended to spend a million dollars constructing an erosion-control structure in the Chaco Wash. This told us the agency knows it has a "functionality" crisis on its hands.

But how can proper functioning condition be restored if the Park Service's hands are tied by a cultural value that says Chaco must be protected from incompatible activities, even those that might have a beneficial role to play in restoring the park to health?

As I drove home, I realized that this tension between "value" and "function" at Chaco was a sign of a new conflict spreading slowly across the West—symbolized by a fence. The cherished "protection" paradigm, embedded in the conservation movement since the days of John Muir, rubbed against something new, something energetic—something beyond the fence.

## BANDELIER NATIONAL MONUMENT
*Near Los Alamos, New Mexico*

The passage of the Wilderness Act in 1964 was a seminal event in the history of the American conservation movement. For the first time, wilderness had a legal status, enabling the designation and the protection of "wildland," which had been under siege in that era of environmental exploitation. Energized, the conservation movement grabbed the wilderness bull by both horns and has not let go to this day. But the act's passage also had an unforeseen consequence—it set in motion the modern struggle between value and function in our Western landscapes.

This tension took a while to develop. In 1964, there was intellectual harmony between the social and ecological arguments for the creation of a federal wilderness system. No reconciliation was necessary between the act's definition of wilderness as a tract of land "untrammeled by man . . . in which man is a visitor who does not remain" and Aldo Leopold's declaration, published in *A Sand County Almanac* fifteen years earlier, that wilderness areas needed protection because they were ecological "base datum of normality" (1949, 196).

Leopold asserted that wilderness was "important as a laboratory for the study of land health," insisting that in many cases, "we literally do not know how good a performance to expect of healthy land unless we have a wild area for comparison with sick ones" (1949, 197). Author Wallace Stegner extended the medical metaphor when he argued that wilderness was "good for our spiritual health even if we never once in ten years set foot in it" (2007, 353).

But a lot has changed in the years since the passage of the Wilderness Act. While most Americans still believe wilderness is necessary for social and mental health, few ecologists now argue that wilderness areas can be considered as "base datum" of ecological health.

For example, in an article ultimately published in the journal *Wild Earth* (but first published in 2000 in *Forest Service Proceedings*) titled "Would Ecological Landscape Restoration Make the Bandelier Wilderness More or Less of a Wilderness?" the authors—including ecologist Craig Allen, who for nearly twenty years has studied Bandelier National Monument, located in north-central New Mexico—state matter-of-factly that "most wilderness areas in the continental United States are not pristine, and ecosystem research has shown that conditions in many are deteriorating" (209).

In their opinion, the Bandelier Wilderness is suffering from "unnatural change" as a result of historic overuse of the area in the late nineteenth and early twentieth centuries—grazing by sheep, principally—which triggered unprecedented change in the park's ecosystems, resulting in degraded and unsustainable conditions. "Similar changes," they write, "have occurred throughout much of the Southwest" (210).

Specifically, soils in Bandelier are "eroding at net rates of about one-half inch per decade. Given soil depths averaging only one to two feet in many areas, there will be loss of entire soil bodies across extensive areas" (211). This is bad because the loss of topsoil, and the resulting loss of water available for plants, impedes the growth of all-important grass cover, thus reducing the incidence of natural and ecologically necessary fires.

The elimination of livestock grazing with the creation of the park in the 1930s was no panacea for Bandelier's functionality crisis, however. Herbivore exclosures established in 1975 show that protection from grazing, by itself, "fails to promote vegetative recovery." Without management intervention, they argue, this human-caused case of accelerated soil erosion will become irreversible: "To a significant degree, the park's biological productivity and cultural resources are literally washing away" (213).

Their summation is provocative: "We have a choice when we know land is 'sick.' We can 'make believe,' to quote Aldo Leopold, that everything will turn out all right if Nature is left to take its course in our unhealthy wildernesses, or we can intervene—adaptively and with humility—to facilitate the healing process" (214).

I believe new knowledge about the condition of the land leaves us no choice: we must intervene. However, this turns a great deal of old conservation thinking on its head.

For instance, Wallace Stegner once wrote, "Wildlife sanctuaries, national seashores and lakeshores, wild and scenic rivers, wilderness areas created under the 1964 Wilderness Act, all represent a strengthening of the decision to hold onto land and manage large sections of the public domain rather than dispose of them *or let them deteriorate*" (1998, 270, emphasis added).

But we have let them deteriorate—as the Buenos Aires, Chaco, and Bandelier examples demonstrate. Whether their deteriorated condition is a result of historical overuse or some more recent activity is not as important as another question: what are we going to do to heal land we know to be sick?

Clearly it's not 1964 anymore. The harmony between value and function in the landscape, including our protected places, has deteriorated along with the topsoil. This functionality crisis raises important questions for all of us. What, for instance, are the long-term prospects for wildlife populations in the West, including keystone predator species, if the ecological integrity of these special places is being compromised at the level of soil, grass, and water? Also, does protection from human activity preclude intervention, and if so, at what cost to ecosystem health? And on a larger scale, how do we protect our parks and wildernesses from the effects of global warming, acid rain, and noxious weed invasion?

Furthermore, the dualism of protected versus unprotected creates a stratification of land quality and land use that bears little relation to land health. As conservationist Charles Little has written, "Leopold insisted on dealing with land whole: the system of soils, waters, animals, and plants that make up a community called 'the land.' But we insist on discriminating. We apply our money and our energy in behalf of protection on a selective basis." He goes on to say, "The idea of a hierarchy in land quality is *the* tenet of the conservation and environmental movement" (10).

Since John Muir's day, the conservation movement has based this hierarchy on the concept of "pristineness"—the degree to which an area of land remains untrammeled by humans. As late as 1964, when not as much was known about ecology or the history of land use, it was still possible to believe in the pristine quality of wilderness as an ecological fact, as Leopold did. Today, however, pristineness must be acknowledged to be a value, something that exists mostly in the eye of the beholder.

Biologist Peter Raven puts it in blunt ecological terms: "There is not a square centimeter anywhere on earth, whether it is in the middle of the Amazon basin or the center of the Greenland ice cap, that does not receive every minute some molecules of a substance made by human beings" (quoted in Shabecoff 2001, 14).

I believe the new criterion should be *land health*. By assessing land by one standard, a land-health paradigm encourages an egalitarian approach to land quality, thereby reducing conflicts caused by clashing cultural values (theoretically, anyway). By employing land health as the common language to describe the common ground below our feet, we can start fruitful conversations about land use rather than resort to the usual dualisms that have dominated the conservation movement for decades. We can also gain new knowledge about the condition of a stretch of land, and that knowledge can help us make informed decisions. Furthermore, the healthier land is, the wilder it becomes. Nature is all about regeneration, not degeneration. Healthy, diverse, properly functioning, biological-rich land—above ground as well as below—regenerates life in all its wildness. Depleted, eroded, and biologically poor land does the opposite. In the twenty-first century, it may be difficult to achieve ideal conditions for land health, especially in prolonged droughts, but we can certainly move in that direction—toward the relative wild.

For example, I know a chunk of Bureau of Land Management (BLM) land west of Taos, New Mexico, that will never be a wilderness area, national park, or wildlife refuge. It is modest land, mostly flat, covered with sage, and very dry. In its modesty, however, it is typical of millions of acres of public land across the West. It is typical in another way too—it exists in a degraded ecological condition, the result of historic overgrazing and modern neglect. A recent qualitative land health assessment revealed its poor condition in stark terms (lots of bare soil, many signs of erosion, and a lack of plant diversity), confronting us with the knowledge that more than forty years of total rest from livestock grazing had not healed the land. Some of it, in fact, teetered on an ecological threshold, threatening to transition to a deeper degraded state.

Fortunately, as humble and unhealthy as this land is, it is not unloved. The wildlife like it, of course, but so do the owners of the private land intermingled with the BLM land, some of whom built homes there. The area's two new ranchers also have great affection for this unassuming land and want to see it healed.

These ranchers are using cattle as agents of ecological restoration. Through the effect of carefully controlled herding, they intend to trample the sage and bare soil, much of which is capped solid (without a cover of grass or litter, soil will often "cap," or seal when exposed to pounding rain, thus preventing seed germination), so that native grasses can get reestablished again.

Using cattle as agents of ecological restoration is not as novel as it may sound. In fact, in his 1933 classic book *Game Management*, Aldo Leopold wrote more generally that wildlife "can be restored with the same tools that have hithertofore destroyed it: fire, ax, cow, gun, and plow" (1986, xxxi). The difference, of course, is the management of the tool, as well as the goals of the tool user.

I believe conservationists should share the same goal as these ranchers: transform red to green on maps such as that of the Altar Valley and the land west of Taos. Whether we use cattle or some other method of restoration, the result must be a thousand acts of healing, starting at the level of soil, grass, and water. And healing must extend to communities of people as well, both urban and rural. Restoration jobs could be a boon to local economies, and volunteers from environmental groups could help. Turning red to green could unite us no matter what our values.

By developing a common language to describe the common ground below our feet, by working collaboratively to heal land and restore rural economies, by monitoring our progress scientifically, and by linking "function" to "value" in a constructive manner, a land-health paradigm can steer us toward fulfilling Wallace Stegner's famous dream of creating a "society to match the scenery."

## *References*

Leopold, Aldo. "The Ecological Conscience [1947]." In *The River of the Mother of God and Other Essays by Aldo Leopold*, edited by Susan Flader and J. Baird Callicott, 338–46. Madison: University of Wisconsin Press, 1991.

———. *Game Management*. Reprint. Madison: University of Wisconsin Press, 1986.

———. "The Land-Health Concept and Conservation [1946]." In *For the Health of the Land*, edited by J. Baird Callicott and Eric T. Freyfogle, 218–26. Washington, DC: Island Press, 1999.

———. *A Sand County Almanac and Sketches Here and There*. New York: Oxford University Press, 1949.

Little, Charles. "In a Landscape of Hope." In *The World of Wilderness: Essays on the Power and Purpose of Wild Country*, edited by T. H. Watkins and Patricia Byrnes, 1–17. Niwot, CO: Wilderness Society, 1995.

Shabecoff, Philip. *Earth Rising: American Environmentalism in the 21st Century*. Washington, DC: Island Press, 2001.

Stegner, Wallace. "Land: America's History Teacher." In *Marking the Sparrow's Fall: The Making of the American West*, edited by Page Stegner, 260–80. New York: Holt Paperbacks, 1998.

———. "To David Pesonen, December 3, 1960." In *The Selected Letters of Wallace Stegner*, edited by Page Stegner, 352–57. Berkeley, CA: Counterpoint, 2007.

Sydoriak, C., C. Allen, and B. Jacobs. "Would Ecological Landscape Restoration Make the Bandelier Wilderness More or Less of a Wilderness?" *USDA Forest Service Proceedings* 5 (2000): 209–15.

# 10

## THE HUMMINGBIRD AND THE REDCAP

*Devon G. Peña*

There is no indigenizing capitalism. —Kam'ayaam/Chachim'multhnii, "Red Skin, White Masks: A Review" (2014, 190)

My first chance meeting with a hummingbird was in 1987. I was backpacking through subalpine forest below Fluted Peak in the Sangre de Cristo Mountains. It was my first hike in this stretch of high country that looms over Colorado's San Luis Valley. I was dressed in jeans, T-shirt and hoodie, leather hiking boots, and a red "Marx" baseball cap for shade. I was approaching timberline on the final switchbacks before the trail opens onto lush meadows inhabited by cushion plants and willow congregations circling the small alpine lake at the foot of the Mt. Adams cirque. Suddenly, something started buzzing my red-capped head. I was quite spooked at first, thinking it was a really large bumblebee or angry wasp. I ran, hid, and ducked; I ran again, finally squatting behind what I now recognize was a glacial erratic, a six-foot boulder forged in Crestone conglomerate.

Nothing seemed to relieve me of the "attack," and the creature kept buzzing at my red cap. I finally found the wherewithal to take a swat on what must have been a fifth or sixth approach. I nearly hit "it." The turbulence of the slashing redcap affected the creature's flight, and I noticed it landing on an aspen branch, making it sway slightly, some fifteen feet away. I recognized the creature as a hummingbird. It was gleaming emerald green and violet hues in filtered sunlight as it peeked around and poked at the air with a long, thin, black beak. Our eyes connected for a brief second, and then it buzzed at me again: I just stood there calmly embracing the moment as the *chuparosa* (rose sucker) explored the redcap. Flitting about inches from my nose, I noticed a ruby-red throat. The buzzer finally decided my redcap was neither a source of nourishment nor a threat and flew away, disappearing into the aspen grove.

That moment of mutual recognition marks the first in a series of life-changing epiphanies in the Rockies that shaped my path to being in the world. I finally understood something I had felt before but with a new sense of significance that had previously eluded me: Wildness is the dissolution of the boundaries of the self, accepting an invitation to follow a niche-abiding way of life in gentle relation with more-than-human beings. The niche-abiding life is situated in place. Wherever it emerges, such a life is cognizant of the coupling of culture and nature as defined by the conditions of the self-willing bioregional ecosystem.

I grew up in Laredo, *Tejas*, astride a contested and transgressed borderline, and we seldom saw hummingbirds in mesquite-prickly pear brush country. We did have plenty of large stinging wasps, which might explain my initial upset during that first chance meeting with a Mexican hummer in the Rockies. That ruby-throated *colibrí* taught me that knowledge of place is a necessary precondition for finding one's right livelihood, and this requires the coeval ordering of our relations with more-than-human beings. This is the *relationality of wildness*.

Perched in willow thickets along the ridge of Culebra Peak, known as *El Comal*,[1] the hummingbird surveys the ribbon-like acequia farms far below and sees that these afford an open invitation to visit and dwell. From the mountain ridge, the hummingbird can assess vast networks of riparian corridors comprising tall trees, willow thickets, and dense flowering vegetation. These are welcoming *refugia* and are shaped by the acequia water webs that transect the bottomlands to feed the fields. The hummingbird knows it will find verdant Timothy hay and alfalfa fields and polyculture *milpas* of heirloom maize, *bolita* bean, and *calabaza*, all set aflame in red, blue, violet, yellow, and white blossoms, the sweet jewels of nectar amid the wild green and ever-shifting mosaic. The hummingbird relishes the vibrant matter unleashed by the incomplete control of water under the intense heat of alpine summer. Viewed from this perch, the acequia heritage landscape is a gift of coevalness that extends the possibility of continued hummingbird flourishing alongside the busy beaver-like human creatures.

The incomplete control of water by acequia irrigators is what produces the lush riparian corridors of the San Luis area. There are more than ten thousand acres of wetlands in the Culebra watershed, and all are a direct result of the subirrigated seep associated with acequia methods. These values remain illegible to the state engineers and hydrologists. Meanwhile, the hummingbirds delight in the floral bounty and organic integrity of our working heritage landscape ecology. Still, our always-morphing farms reside within the wildness of the watershed. This is actually our strength, since informal tradition and custom allow for the continued sharing of water with more-than-human beings. Our wastefulness [*sic*] is a generous gesture to more-than-human beings, and it is this ecological value that remains incomprehensible to water engineers, water judges, and privateers. Our autonomy requires that we act in conformity with acequia law, which is relational and bound to the precept that water is a shared asset-in-place requiring a unique form of local regulation of use rights subject to the norms of mutual aid, cooperative labor, and shared scarcity.

The incomplete control of water in the irrigation common creates the famous "cultural landscape" of the Culebra River acequia villages and their uplands. In the absence of acequias, we would lack the subirrigated Vega village common lands or the riparian corridors of cottonwood, willow, alder, rosehip, and *oshá* communities that transect the bottomlands. These anthropogenic features shelter wild medicinal and food plants, birdlife (including hummingbirds), and even fish and small mammals while protecting farm fields from wind erosion and creating an acequia aesthetic. We would have to do without adobe architecture and the native white flint maize roasted in adobe ovens to produce our renowned *chicos del horno*. Without acequias, the community would lack access to the means of right livelihood and deep environmental knowledge of an arid-sensible way of life. We would lack resilient traditions of local democratic self-governance for cohabitation of the watershed in a manner that includes more-than-human beings in the circle of conviviality.

As a community, we need to recall what the original instructions might pronounce as the first rule of obligation: What works for acequias must work for hummingbirds. The conditions that produce hummingbird niches are the same that create the possibility of a niche-abiding way of life among willing acequia farmers.[2] A tangible material and spiritual connection may emerge that binds watershed biophysical conditions to the *sociocultural performance of place* by local human members of the mixed community.

The multigenerational acequia farmer and ethnophilosopher Reyes García introduced me to the Colorado acequias in 1985 at his farm outside Antonito in the Conejos Land Grant. I think about Reyes every time I look out at the Acequia Institute's riparian long lot in the San Acacio bottomlands and feel the same amazement he expressed at the diversity of life created by acequia flood-irrigated methods. The acequias maximize diversity by encouraging the mosaic-shaping processes of riparian wildness itself, resulting in the distinct morphology of these farmlands. The self-willing watershed brings floods and storms that alter conditions in the acequiahood; windstorms deposit soil from the mesa tops; new associations of plants and trees emerge. The acequiahood extends across the land as an extension of the effects of the immanent power of the watershed's own "agency." In other words, the watershed created the conditions that today support our farms through successive ancient depositions of glacial outwash, the gradual formation of alluvial fans, and the accumulation of windswept grains from ancient and now fossilized sand dune fields that are the trifecta of our most productive soils.

The best practitioners of flood irrigation regenerate the land because the water webs capture sediment suspended from the mountain cirques and parks all the way down to our fields, where the irrigator-artist can gently deposit it. This induces a coeval biophysical process that replenishes soil horizons and diverse habitat niches. Acequias extend a primary force of life—gravity—as it drives icy waters from deposits of melting snowpack through the dendritic webs *to assert the presence of wildness* in the vibrant materiality of diverse species scattered in niches across cultivated fields, orchards, riparian strips, wetlands, woodlands, meadows, and the elegant polycultural home kitchen gardens of the farmers inhabiting the shifting mosaic.

There is a decolonial orientation for being in the world that I will call *unpossessive presence*, and it can serve as an adopted trait wherever a respectfully situated human person negotiates "going native." Such a state of being, for the nonnative interloper, is often as fleeting as hummingbird flight, so he or she must learn to "nest" within a more enduring connection to a whole that is truly greater than the sum of its parts.

I believe Aldo Leopold, who gave us an outline of the "land ethic," was singular among Western thinkers in arguing with clarity that wildness is a quality of the land as *self-willing* and that its health is determined by its "capacity ... for self-renewal" (1949, 221). He meant this in the sense of a self-organizing superorganism enacting change through evolutionary geomorphological, biochemical, and other long-duration processes that constitute the interconnected pathways of energy and matter, the very forces that create the conditions that make the enactment of life possible. He understood the importance of a niche-abiding life as suggested by his idea that we should be "plain members and citizens" rather than "conquerors" of the land community and some all-too-brief musings on the shamefully commoditized view of the land among the world's three major Abrahamic traditions.

The niche-abiding life is situated in place. Wherever it emerges, such a life is cognizant of the coupling of culture and nature as defined by the conditions of the self-willing bioregional ecosystem. Transcending the universalizing or totalizing human desire to reign supreme (sovereign) over others is a foundational virtue of a niche-abiding life, along with respect for the "pluriversality" of place-based ways of being. Among indigenous peoples, the idea of the niche is described by what ethnobiologist Enrique Salmón calls "kincentric landscapes"—learning the place in which one lives by being in relation to all life forms and especially more-than-human beings (21–22).

White settler colonial projects present severe challenges to the niche-abiding life. In a recent book that "indigenizes" Karl Marx, Glen Coulthard (2015) explores the concept of primitive accumulation—the forcible and violent separation of the immediate producer from the means of production—as the quintessential form of colonial dispossession. He interprets the political, economic, and juridical strategies deployed by invasive forces as colonial, rather than simply capitalist, relations aiming to dispossess First Peoples of their aboriginal territories rather than reducing them to abstract labor.[3] My own sense is that this reveals further how Western settler colonial projects are compelled by a pathology born of the colonizer's desire for dominance beyond the niche.

We are not ghosts of the primitive accumulation, of a forced separation from the land: There are living and working heritage landscapes across the planet that defy "end of place" scenarios. It is within the cycles of epochal Indigenous resistance to capitalist colonial enclosures that the struggle for equality among groups, across species, and toward future generations starts to make ethical and political sense. This requires relational solidarity with self-willing land+water as vibrant coactants and shapers of the world. We must "de-occupy" all colonized spaces, rupture and interrupt the illusionary spell of the commodity relation, enact autonomous collectivities whose very existence negates the self-signifying empire of capital that assembles itself from endless cycles of epistemic violence that suppress other ways of knowing and being in the world.

The practical ethno-philosophical response to these challenges is something I have learned about during three decades as a student and practitioner of acequia agroecology in the San Luis Valley. Our community acequia (irrigation ditch associations) are renowned as an institution of civil society that allows the human being (as irrigator) to find a niche amid a whole that is greater than the sum of its parts. The interwoven hydraulic, agroecological, and sociocultural systems of community acequias in Colorado's Culebra watershed appear to follow a fundamental niche-abiding rule: The irrigator exercises incomplete control of water and in the process contributes to the well-being of more-than-human beings. The Islamic water law celebrated in Córdoba, Al-Andalus (Spain), included the *right of thirst*, which established the edict that any living thing with thirst has a right to water, and denying water to those with thirst is "sinful."

Yet the sin of primitive accumulation continues to deny water to the thirsty. For example, in San Luis, citizens of the acequiahood have struggled against clear-cutting of the forests on La Sierra common lands because the trees are deemed necessary to a stable snowpack and acequia irrigation supply; we have resisted mining on the mountains because this threatens our water quality; and we have opposed corporate monoculture farms that tap the aquifers and hence undercut the quantity of supply to acequias. Industrial-scale logging, mining, and agricultural monocultures have all unsettled the central role of humans as a keystone species democratically sharing the water in a way that is niche-abiding with the more-than-human world.

As a scholar of indigenous cultures and forms of autonomous self-governance, I have yet to hear of a single culture that has a word for "individual" among the thousands of indigenous languages of Turtle Island First Peoples. There are words that translate into "human being" and "person," as well as various words for relatives (mother, father, brother, sister, and so on), and these may also refer to more-than-human beings. But there is no word that translates directly as "individual." The word *individual* does not even become part of the common linguistic parlance in a Western context until the 1860s in the United States, when it began to serve as the interpretive linchpin of legal and juridical discourses pertaining to the American classical liberal notion of the autonomy of the human being in relation to politically constituted state formation.[4] This discourse grants that under the US Constitution legal standing is assigned *only to individuals*, including especially the right to own property; this stands even as *Citizens United* now envisions "personhood" for corporations. Sociocultural and bioregional groups, mutual aid collectives, and multispecies amalgamations and associations in local and bioregional environments—none of these truly exist as "rights holders" in the aftermath of the epistemic violence unleashed by the Republic of Property and its narrowly misconstrued notion of individual rights (Hardt and Negri 2009). A single hermeneutic sleight of juridical logic allowed the settler colonists to create a constitutional order that violently erased other ways of being in the world.

A shared concept of "being" in Mesoamerican and Native American ethnophilosophy is that a person becomes human only in relation to others—*all* others. Otherwise, we are just "skin."[5] Chicana/o scholars frequently explain Mesoamerican epistemology by enunciating the Mayan ontological dictate, *In Lak'ech*: "You are my other self." This *alter*Native epistemic enunciation dispenses with the desire to serve as "mirroring others" (enanthiomorphs): This is not the identity politics of (mis)representation but rather the decolonial enunciation of a niche-abiding life that requires no discourse with or recognition from the colonizer, as the struggles against logging and mining in San Luis attest.

The absence of a concept of individual rights is understandable once we examine shared principles of the political morality of Turtle Island First Peoples (indigenous inhabitants of the North American continent). One shared principle is the idea that the first rights granted to human beings involve the delegation of duties for the fulfillment of obligations to the land and place as shared home spaces (niches). These obligations are morally grounded in the self-willing agency of the land+water and derive from what Melissa Nelson and her colleagues call "Original Instructions" (Nelson 2008). The people may follow these instructions to inhabit places without being dominant or abusive. The land+water is the "first" or "original" subject with the overarching power to grant other subjects their assigned obligations to fulfill community mutual reliance as distinct from individualistic utilitarian interests. A central theme in the political morality of First Peoples is therefore the matter of the fulfillment of collective obligations that grow out of recognition and respect for the boundaries, limits, and interconnective tissues and metabolism of the niche. It is thus deemed essential to sustain modest, nonacquisitive livelihoods, and this ends the colonial capitalist binary of economy against ecology. The norm of modest inhabitation helps us avoid the second contradiction of capitalism, which is the tendency to destroy the social and ecological conditions of its own (and by extension, our) existence. This rule against self-aggrandizement can be enforced through the ethics of *vergüenza*, or "shame," as in the case of many acequia communities. This is a complex moral norm, but basically it establishes an edict to restrain those who do not abide by other niches and are governed by greed and selfishness—precisely the qualities defended by settler colonial societies through their imposition of a juridical regime based on individual rights (Hicks and Peña 2003; Peña 2005; Taylor 2006).

The defense of wildness requires *relational solidarity* because the scale of the destructive force of the Capitalocene demands collective action toward emplacement of alterNative worlds here, now. I am humbly observing how an increasing number of indigenous and other multigenerational place-based communities like the Culebra acequiahood are choosing to withdraw from the neoliberal capitalist state by creating *alter-economies*—postcapitalist commonwealths built on the foundation of antecedent resurgent principles of conviviality. The niche-abiding life strives to create a *solidarity* economy rather than a *predatory* one. I recognize our acequiahood is one limited and always vulnerable constitutive democracy, but we can help change the world as part of a network of bioregional archipelagos of *autonomía* already nestled in valleys across the hummingbird headwaters of the Río Arriba (Upper Rio Grande).

According to Alexandre Kojève, "No animal can be a snob" (quoted in Agamben 2004, 9). Perhaps this statement is meant to reveal the folly of the *possessiveness of being* that some philosophers cling to in order to articulate their exceptionalist views of human culture and history. The snobbish human thrives in a myopic hierarchy that relies on formalized rulemaking to establish some distance between the gains of selfish interest and the violence this unleashes on others. The Western doctrine of prior appropriation is an example of a juridical order beholden to such rigid snobbishness. The problem in Colorado is that the modern juridical order misjudges the incomplete control of water in acequias as a case of "economic inefficiency." The law then appeals to the colonial settler logic and anthropocentric metric that splits "beneficial" from "nonbeneficial" uses of water—phreataphytes, the deep-rooted plants that thrive along riparian corridors, and mutual reliance interests notwithstanding.

The niche-abiding life eschews snobbishness. This includes the methodological individualism of rational choice theorists who believe they have tapped the secret of human nature: We are all declared greedy, self-interested, utilitarian individuals. The core belief is that the freedom to perform our acquisitiveness and self-aggrandizement will perfect the market's ability to maximize the social good. These snobbish colonizer voices have never provided a more profoundly troubling illustration of pathological ethnocentrism and negative ontological logic. It should hardly beckon anyone to celebrate a peculiarly Euro-Anglo-American notion of civilization (or culture) as separation from wildness, a concept bound to the most severe capitalist rationality comprising an *ideology of disconnection*, as Michael Taylor (2006) deftly argues.

Looking at the matter of water law in the context of long-standing acequia battles with the State of Colorado over "abandonment" and "condemnation" of our water rights, the prevalent conceptualizations of the territorial enclosures of settler colonialism can be read as dominionist and exceptionalist ontological positions because they rule out the possibility of a politics of *coevalness* among humans, other organisms, and ecosystems (Smith 2011). The anthropocentrism of human sovereignty, understood here as state- and subject-making, over ecology remains extant even in wilderness advocacy circles where this *ruling over* arrives just in time to "save nature" in the name of conservation biology. Presumably, the more business-oriented snobs can arrive later to capitalize on nature's biodiversity, as is currently illustrated by acts of "biopiracy" across the planet.

Any juridical order championing human sovereignty over the environment inevitably reduces other species and organisms, and their collective associations, to the "bare life" by misrecognizing these as nonsubjects stripped of the guarantees of political, social, and legal standing on their own self-determined terms.[6] No subject, including extant ecosystems, can sustain itself as self-willing (or wild) under the spell of the empire of the commodity form or the gaze of a narrowly "scienticized" ecology.

The decolonial indigenous (home)land ethics of a niche-abiding life require that we challenge the biopolitics of the control and management of ecosystems. Biopolitical life manages the coupling of social and ecological subsystems with varied aims in mind, to be sure, but the two dominant frames both suffer from a form of disconnection, since they either reduce nature to natural resources that serve the aims of commodity fetishism or seek to protect wilderness [sic] areas by keeping these essentially separate from humans. Both of these are colonizing projects because they fail to overcome the nature/culture binary and erase the fact that people have already inhabited these self-willing lands, as Native peoples and hummingbirds have done much longer than the advent of colonialism and its discontents.

Thinking back to that first chance meeting with the ruby-throated hummingbird, I see that my response was also snobbish. Don't buzz my redcap! I am better than you! How dare you dart at me! The joust with the hummingbird taught me to relinquish my stiff, privileged position: a sort of "stickiness of being" that blocked openness to this more-than-human being. I had to unlearn *possessive presence* by stepping beyond the realm of a purely human culture. I had to be suspended with the hummingbird, both precarious and vulnerable. We are still locked in a recognizing gaze, swiftly flitting away to challenge the way that holds only antipathy toward all things wild, untamed, and common.

## Notes

1. *Translation:* "Tortilla grill." This is a local place name and is not featured in any maps that I know of. It refers to the flat-topped inclined diagonal that comprises the northwest face of the approach to Culebra Peak as seen from the Valle Creek drainage. The view from the meadows by the old cabins that Jack Taylor built in the 1960s makes this feature looks like a flat grill used to make or reheat tortillas.

2. However, the hyperobject of climate change, delineated by Morton (2013), presents a challenge to the future of acequias through processes that threaten to end the continued ecology of a self-willing land and, locally, may signal the decline of a water democracy based on the generous incomplete control of water. I address the challenge of climate change and acequias in my forthcoming book, *The Last Common* (2016).

3. I am summarizing from the advance Kindle e-book copy, especially the discussion in chapter 1 under the subhead "Karl Marx, Settler-Colonialism, and Indigenous Dispossession in Post White Paper Canada."

4. I thank my colleague Stevan Harrell for offering this observation during a guest lecture I presented to the environmental anthropology class at the University of Washington in autumn 2013.

5. This is the case in Nahuatl linguistics, according to Tezozomoc; personal communication to author (June 2007).

6. I have drawn throughout this essay from the work of Giorgio Agamben on the "bare life" (1998); the anthropological machine and the separation of (some) human beings from their animal relatives (2004); and the "state of exception" (2005). I have also consulted Smith (2011) on culture/nature dichotomies and biopolitics.

## References

Agamben, Giorgio. *Homo sacer: Sovereign Power and the Bare Life.* Stanford, CA: Stanford University Press, 1998.

———. *The Open: Man and Animal.* Stanford, CA: Stanford University Press, 2004.

———. *State of Exception.* Chicago: University of Chicago Press, 2005.

Bennett, Jane. *Vibrant Matter: A Political Ecology of Things.* Durham, NC: Duke University Press, 2010.

Coulthard, Glenn. *Red Skin, White Masks: Rejecting the Colonial Politics of Recognition.* Minneapolis: University of Minnesota Press, 2014.

Hardt, Michael, and Antonio Negri. *Commonwealth.* Cambridge, MA: Harvard University Press, 2009.

Hicks, Gregory, and Devon Peña. "Community Acequias in Colorado's Rio Culebra Watershed: A Customary Commons in the Domain of Prior Appropriation." *University of Colorado Law Review* 74 (2003): 387–486.

Kam'ayaam/Chachim'multhnii (Cliff Atleo Jr.). "Red Skin, White Masks: A Review." *Decolonization: Indigeneity, Education and Society* 3, no. 2 (2014): 187–94.

Leopold, Aldo. *A Sand County Almanac and Sketches Here and There.* New York: Oxford University Press, 1949.

Morton, Timothy. *Hyperobjects: Philosophy and Ecology after the End of the World.* Minneapolis: University of Minnesota Press, 2013.

Nelson, Melissa K., ed. *Original Instructions: Indigenous Teachings for a Sustainable World.* Rochester, VT: Bear, 2008.

Peña, Devon. *Mexican Americans and the Environment: Tierra y Vida.* Tucson: University of Arizona Press, 2005.

Salmón, Enrique. *Eating the Landscape: American Indian Stories of Food, Identity, and Resilience.* Tucson: University of Arizona Press, 2011.

Smith, Mick. *Against Ecological Sovereignty: Ethics, Biopolitics, and Saving the Natural World.* Minneapolis: University of Minnesota Press, 2011.

Taylor, Michael. *Rationality and the Ideology of Disconnection.* Cambridge, UK: Cambridge University Press, 2006.

# 11

## LOSING WILDNESS FOR THE SAKE OF WILDERNESS

### The Removal of Drakes Bay Oyster Company

*Laura Alice Watt*

On January 3, 2015, a rousing wake was held in the town of Point Reyes Station in Northern California. Bluegrass music played, and a large, noisy crowd enjoyed a plentiful barbeque. I was not there, but I imagine a certain hollowness: people delighting in the winter sunshine and sharing memories but also sorrowful in their recognition of the end of an era. This gathering was to mark the passing not of an individual but of an eighty-two-year-old oyster farm, which operated on state-controlled leases over the tidelands in Drakes Estero, a glove-shaped estuary at the heart of Point Reyes National Seashore. The oyster farm was forced to close by the Secretary of Interior in a decision made in November 2012. The accompanying press release cited National Park Service (NPS) law and policy concerning commercial operations and wilderness and represented the culmination of five years of fierce, polarizing debate that pitted advocates for wilderness against those for sustainable agriculture. Lawsuits kept the company, Drakes Bay Oyster Company (DBOC), in limbo for several years, but by the end of 2014, the doors were shuttered and the last oysters were harvested from the estuary.

The controversy drew national attention, as it was not the usual industry-versus-nature debate. On the one hand, national environmental organizations sought official designation of a marine wilderness to be enjoyed by hikers and kayakers. On the other, the oyster farm operators and local foods advocates insisted that the historic operation, in place since the 1930s, was doing no harm and should be allowed to continue as an example of sustainable agriculture. Drakes Estero was already managed by the NPS as wilderness, with the lone exception of allowing the oyster boats access to their racks. The fact that the Estero has been routinely cited for decades as one of the most pristine estuaries on the Pacific Coast, despite the more than eighty-year presence of a continuously operating oyster farm, suggested that cultivating oysters is more compatible with wilderness qualities than many environmental advocates have been willing to admit.

I became involved with this controversy in 2007 after two environmental advocates published an op-ed in a local newspaper, arguing that DBOC was required by wilderness law to close in 2012, when its reservation of use and occupancy was due to expire. They asserted that by designating the Estero as "potential wilderness," Congress had intended that all inconsistent activities and uses be removed, and thus the oyster operation was the sole remaining obstacle to wilderness protection. My doctoral research on the management of Point Reyes meant I was deeply familiar with the legislative history of the wilderness designation, and I wrote a response asserting that the environmentalists' argument was a distortion of congressional intent.

While I did not know the Lunnys, owners of DBOC, at the time I wrote that first op-ed (they since have become good friends), I felt that the NPS and environmental groups' treatment of their family-run business and employees was unfair. The argument for their company's ouster relied on not only what turned out to be shoddy and inaccurate science—a review by the National Academy of Sciences in 2009 found that the NPS had "exaggerated the negative and overlooked potentially beneficial effects of the oyster culture operation" (3)—but also a very selective reading of history through the lens of an overly idealized vision of wilderness that has taken hold among environmental activists since the 1990s. In contrast, the idea of the *relative* wild is actually rooted in the pragmatic approach taken with the original 1964 Wilderness Act, and when we stray from that, into more purist versions of what wilderness "ought" to be, we lose the possibility and power of that relativeness, that humans may have beneficial roles to play in nondomesticated lands and waters.

## A Pragmatic Wilderness Law

At first glance, the passage of the 1964 Wilderness Act, which defined wilderness as "untrammeled," a landscape "retaining its primeval character and influence," would seem to be the culmination of what environmental historian William Cronon has described as the "unexamined foundation on which so many of the quasi-religious values of modern environmentalism rest" (1995, 80). The act compelled federal land management agencies—Forest Service, Park Service, BLM, and Fish and Wildlife Service—to identify roadless areas of their holdings larger than five thousand acres in size with opportunities for solitude and primitive or unconfined types of recreation so that Congress might formally designate them as wilderness areas. Commercial enterprises and permanent roads were prohibited, along with structures and motorized or mechanical uses.

Yet the modern concept of wilderness, particularly as formalized in the 1964 act, must be understood as part of a broader cultural shift from a producer to a consumer society. The rapid proliferation of the automobile in American society in the early twentieth century—from only eight thousand registered in 1900 to more than twenty-three million by 1929—and federal support for road building profoundly expanded the reach of the car, and the traveling public with it. Early NPS employees and supporters embraced these trends, specifically designing parks to accommodate cars and car campers, but others involved with public lands management, including a young timber surveyor for the US Forest Service named Aldo Leopold, saw this welcoming stance toward vehicles as threatening something essential about the wild. It was in the context of this burst of road building and recreational development of America's public lands that Leopold first suggested the need for wilderness preservation (Sutter 2002).

Leopold and his fellow founders of the Wilderness Society, the organization most instrumental to the eventual passage of the 1964 Wilderness Act, made important exceptions to their proposed policy in order to accommodate commercial enterprises, particularly subsistence or small-scale local land uses. For example, the Gila Wilderness Area, first protected by a plan written by Leopold himself in 1924, allowed existing grazing operations to remain while prohibiting large-scale logging (Meine 1988). He also wanted to reconnect aesthetic appreciation to land use rather than create exclusive scenic preserves: "By sequestering natural beauty from economic use, Americans were doing themselves a disservice, Leopold thought" (Sutter 2002, 94). During his work in the Southwest, Leopold helped shape the earliest Forest Service policies in the 1920s for locating roadless areas, protecting them from road building and other forms of development, and recognizing wilderness recreation as a dominant use. These policies allowed limited logging and grazing to continue, provided they did not require new roads. Formalized in 1929 as Regulation L-20, this policy created the first system of protected wilderness and labeled the areas as "primitive," emphasizing their less-developed nature while acknowledging that many of these places were neither pristine nor untouched.

After World War II, though, the Forest Service managed its lands ever more intensely for timber production, using increasingly mechanized technologies to build logging roads into more difficult terrain and onto steeper slopes. At the same time, the NPS was going through a development phase of its own, articulated most clearly by the Mission 66 campaign, a ten-year effort begun in 1956 to upgrade facilities, expand services, and particularly improve and add to its road system (Harvey 2005). In addition to threats of proposed dams in places such as Dinosaur National Monument, this intensification of road development across many western public lands, in service of both timber extraction and tourism, drove much of the push to establish a national wilderness protection law. Yet the purpose was not to "lock up" the public lands; as historian Jay Turner writes, "Even in its earliest drafts, the Wilderness Act included provisions to meet the concerns of rural Americans and the mining, timber, and grazing industries" (30). The bill was designed more to prevent *new* disruptions of existing wilderness than to aggressively push out long-established local uses.

The resulting law includes lyrical language, penned by Wilderness Society executive director Howard Zahniser, stating, "A wilderness, in contrast with those areas where man and his works dominate the landscape, is hereby recognized as an area where the earth and its community of life are untrammeled by man, where man himself is a visitor who does not remain." Yet section 4(a) of the 1964 Wilderness Act clearly states that wilderness designation is *supplemental* to the purposes for which national parks, forests, and wildlife refuges are established and administered. Also, section 4(d) lists the special provisions of established uses within wilderness areas, including grazing and motorboat usage, and other preexisting private rights and uses. Most importantly, the 1964 Wilderness Act is structured like a zoning law, acting primarily as a restraint on future federal agencies' actions, rather than limiting existing private actions. It was a pragmatic response to prevent overdevelopment of public lands by the very agencies that managed them.

## Working Wildness in the Point Reyes Wilderness

Point Reyes is a place where a portion of the landscape has been "rewilded," with more than twenty-five thousand acres of formally designated wilderness, plus roughly eight thousand acres of "potential" wilderness, almost all of which are tidelands. At the seashore, as elsewhere, the unspoken goal of this area's management has been to untrammel—to deliberately create a scene of seemingly pristine nature. The maintenance of such a scene has produced a tension between actions that reduce or eliminate traces of human land use in the rewilded parts of the park and actions that continue human uses in the working portion of the park (Watt 2002). But including the tidal areas in wilderness was intended to limit the NPS's ability to develop recreational services, and the new category of "potential wilderness," first used in the 1976 bill designating land at Point Reyes and several other parks, allowed Congress to designate areas the federal government did not fully own; it was not aimed at removing existing land uses. The contradictions inherent in this category set the stage for continued disagreement over what *wilderness* means, ultimately building to a heated standoff over Drakes Estero and the discontinuation of the oyster farm.

Many wilderness advocates in the recent oyster farm controversy asserted that by including Drakes Estero in the "potential wilderness" category, Congress intended the oyster farm to cease operation in 2012 once its existing reservation of use expired. This was often described as a "promise," the breaking of which could threaten the entire national wilderness preservation system. Yet in my own research, reading through everything I've been able to find about the designation of wilderness at Point Reyes in the 1970s—planning documents, comment letters from environmental organizations and members of the public, and testimony from congressional hearings, as well as the formal bills and reports and subsequent management plans—I have not come across *any* statements anticipating closure of the oyster farm in 2012.

In contrast, quite a number of statements suggest the opposite: that the oyster farm was intended to continue under potential wilderness designation, with no clear end point or expiration date. For instance, in the 1974 *Final EIS for Proposed Wilderness*, the NPS wrote, "This is the only oyster farm in the seashore. Control of the lease from the California Department of Fish and Game, with presumed renewal indefinitely, is within the rights reserved by the State on these submerged lands . . . and there is no foreseeable termination of this condition." Throughout the legislative history of the Point Reyes wilderness area, there is extensive discussion of the areas that should or should not be included in the wilderness designation, and wherever there is mention of the oyster farm, the tone is unambiguous: It should be allowed to continue operation. There is no hedging of "until its reservation runs out," nor any setting of a specific deadline—the language is simple and clear. The pragmatic view of wilderness that held sway at the time could easily accommodate this then-fifty-year-old relationship with the estuary.

Furthermore, if a permit *had* been issued in 2012 for DBOC to continue production, the Estero's potential wilderness designation would remain unchanged, as would the park's management of the estuary. There is nothing about allowing a prior nonconforming use to continue in a potential wilderness area that "rolls back" or eliminates congressional designation. There are other examples of commercial operations utilizing lands designated as potential wilderness within national parks, such as the High Sierra camps in Yosemite, which have operated commercially for years and have an enormous waiting list for registrants; nothing about the camps' presence has compromised the potential wilderness status of those lands.

It is important to remember that the 1976 wilderness bill was the *first* congressional use of the potential wilderness category. There is still no formal definition of *potential wilderness* by Congress that requires nonconforming uses to cease within any set time frame or process. In 2007, the same year that the DBOC controversy exploded, an NPS official testified before Congress that Southern California Edison could continue its use of a check dam for hydroelectric power within proposed potential wilderness in Sequoia-Kings Canyon "as long as it wants," with no hints toward removal over any time frame.[1] "Potential wilderness," when one considers the historical record, was not conceived of as a way to gradually phase out nonconforming uses; it was a way to *accommodate* them, allow the rest of the area to be managed as wilderness, and simplify the administrative process of declaring those acres to be part of the "full" wilderness if/when the nonconforming uses cease.

Furthermore, at Point Reyes at least, existing nonconforming uses were often considered compatible with wilderness rather than requiring the lesser "potential" designation. Instead of aiming to remove an existing use, wilderness at the seashore was intended to prevent further development by the NPS. Reading through the 1976 wilderness hearings, it appears that the primary motivation for environmental advocates' insistence on wilderness status for the tidal areas was to exclude motorized vehicles. The Sierra Club's representative testified, "The possibility of jeeps and motorcycles having access to the Estero shore and adjoining area is a frightening one." Similarly, a letter from the Marin Conservation League specified, "MCL strongly urges inclusion in Wilderness of the quarter-mile strip of tidelands and Drake's Estero. The fragile and important estero must have protection from recreational motorboats. The beaches must be protected from off-road vehicles . . . and we do not object to the non-conforming use of the Johnson Oyster Co. operation in Drake's Estero." The hearings wrapped up with letters from California assemblyman Michael Wornum (9th District) and Jerry Friedman (then serving as chairman of the Marin County Planning Commission and representing numerous local organizations) that specifically referenced oyster farming at Drakes Estero as a nonconforming use that should continue under full wilderness designation.[2]

In response, the NPS insisted that the tidelands could not be designated as wilderness, not because of the oyster farm, but due to incomplete federal title; the State of California still held fishing and mineral rights to these lands. The compromise, brokered in the eleventh hour by Representative John Burton, was to categorize 8,003 acres, mostly the submerged tidelands, as *potential* wilderness.[3] The NPS's representative, Dr. Richard Curry, agreed with this approach, stating that the NPS had no objection to the areas where the state had retained rights being designated as potential wilderness. In his written statement, he specified that the NPS would now recommend tidelands as potential wilderness, "to become wilderness when all property rights are federal, and the areas are subject to National Park Service control."[4] This compromise was hailed in the local newspaper the following week, quoting Burton: "This will allow the ownership question to be resolved in the future" (*Point Reyes Light* 1976). It was a question of legal title, not of commercial use, that resulted in Drakes Estero being designated as potential wilderness.

## *Losing Wildness for the Sake of Wilderness*

In contrast to the pragmatism of the earlier years of wilderness legislation, the 1980s and 1990s brought a new vision to the forefront of the environmental movement: a far more absolute, purity-based ideal of wilderness as perfectly pristine. Groups like Earth First! and the Wildlands Project demanded wilderness proposals based on the new science of conservation biology and a new approach to wilderness recreation exemplified by "leave no trace." This new ideal, envisioned as pristine, unpeopled, and ahistoric, "left little room for hunters and rural westerners, who had their own ways of appreciating wilderness" (in Turner 2007, 256). It also began to reorient the mainstream environmental organizations' goals, as groups like the Wilderness Society and Sierra Club increasingly adopted this more fundamentalist conception of idealized wilderness. At Point Reyes, representatives of the National Parks and Conservation Association and Sierra Club, as well as the locally based Environmental Action Committee of West Marin, employed this pristine ideal as a crowbar with which to pry the oyster farm out of the park landscape, warning that allowing DBOC to continue past the 2012 expiration of its reservation of use would "overturn" the Estero's wilderness designation.[5]

Yet the wilderness status at Point Reyes was never actually in danger: Drakes Estero's designation as potential wilderness meant that the estuary had been *managed as wilderness* ever since, with the sole nonconforming elements of maintaining the oyster rack structures, which long predated the designation (and the park itself), and the use of an outboard motor to propel the harvest boat. While the Wilderness Act prohibits commercial operations in wilderness areas, DBOC's "commercial operation" itself was on shore, on land that is historically part of the seashore's pastoral zone and that was not included in the wilderness designation. DBOC was part of a long history of fishing and mariculture in West Marin, and many visitors have maintained traditions of hiking the estero or kayaking its waters and then gathering around a picnic table to celebrate with a plateful of oysters. For them, there was no either/or between sustainable agriculture and the wild.

As a point of comparison, there was little difference between DBOC's use of the potential wilderness of the estuary and that of commercial kayak tours. Both provided a service to paying customers that relied on the natural ecosystem, without inputs or manufacturing. Oysters thrive in the same wild conditions that other organisms do, and DBOC workers guided people's use of the estuary's shellfish resource in a similar way that a kayak guide shows visitors to the most spectacular spots. DBOC's on-shore shop hosted as many as fifty thousand visitors a year, making it a major recreation destination, particularly as local food increasingly became a tourist draw for the West Marin area. An oyster even tastes wild, bringing the sharp brininess of the sea to our mouths along with a deep appreciation of place, like the idea of *terroir* in winemaking.

A more direct parallel exists between the oyster operation and livestock grazing, a practice specifically grandfathered in, with appropriate restrictions and oversight, by the 1964 Wilderness Act. The cattle or sheep move out across the landscape, finding their own forage in the meadows of a wide variety of public lands, and convert something inedible to humans—grass—into useful meat and fiber. Human guardians lightly guide their movements, but the animals do the work themselves. Similarly, the oysters hang around in the water column and filter out the nutrients they need, turning something inaccessible to us into (delicious) edible protein—the oyster racks and bags are akin to fences and the oyster workers more like cowboys who check on the bivalves' progress and safety and then occasionally round them up, without diminishing the wild ecosystem on which they depend.

Furthermore, a strong sense of reciprocity existed between the oyster operation and its surroundings, in a relationship that transcended one-way extraction. While few reliable studies have been done on their role in the Drakes Estero ecosystem, oysters are generally recognized in the scientific literature as being beneficial to their ecosystems, maintaining or restoring water quality and providing habitat to other species. Up and down the Atlantic Coast of the United States, oyster cultivation is currently being employed to help restore damaged estuaries and local maritime economies. From Drakes Estero, DBOC sustainably produced roughly 40 percent of California's oysters, all sold exclusively within the San Francisco Bay Area; importing an equivalent quantity of this high-demand product from other locations, such as Seattle, Chesapeake Bay, or Japan, comes with a greatly increased carbon footprint. In addition, oyster workers were "eyes on the water" for recreationists on the estuary and frequently helped or even rescued kayakers stuck in the mud or having other trouble. In recent years, DBOC donated many tons of oyster shell to other Bay Area restoration efforts—including projects to restore native oysters to the Bay and to improve nesting habitat and success rates for the threatened western snowy plover—further helping to cocreate wildness in the region. Since only oysters shucked on-site contributed to this stockpile of oyster shell, and DBOC was California's last operating oyster cannery (all other oyster farms only sell oysters in their shells, rather than shucked and packed in jars), this resource has disappeared along with the company.

The wildness of the estero and the passive production of oysters are not as mutually exclusive as environmentalist rhetoric sometimes suggests. I would argue that Point Reyes represents the future for much of parks management, as we will increasingly need to reconcile the presence of humans, both in the past and looking forward, in the natural world, finding new ways for them to coexist and complement one another. As stated by Cronon, "If we wish to preserve wild nature, then we must permit ourselves to imagine a way of living in nature that can use and protect it at the same time" (1996, 55). Because of this, I personally believe that DBOC should have been embraced as an integral part of our experience of Point Reyes, as an example of relying on nature for something in addition to recreation. It could have underlined for visitors the interdependence that has always existed on the peninsula between nature and humans. I am not the only one to make this suggestion: in a letter to Senator Dianne Feinstein in October 2012, author Michael Pollan wrote, "There are deep roots to the hostility of environmentalism toward agriculture... An 'all or nothing' ethic that pits man against nature, wilderness against agriculture, may be useful in some places, under some circumstances, but surely not in this place at this time." The Lunnys' oyster farm, he wrote, "stands as a model for how we might heal these divisions."

But instead, DBOC has closed. A local newspaper reported less than three months later, "On the windswept shore of Drakes Estero, it's as if 80 years of oyster farming have been erased overnight" (Kovner 2015). The former location of the oyster sales shack, cannery, and wooden docks has been "scraped clean" by a government contractor. Standing on the shore now, instead of seeing pristine wilderness, I see heavy machinery preparing to remove the decades-old oyster racks. The Lunny family who owned the business has been driven nearly to bankruptcy, and roughly thirty full-time employees — men and women who had worked on the estuary for fifteen, twenty, or even more than twenty-five years — have been put out of work and, for those who also lived on-site, put out of their homes as well. A long tradition of cultivation has vanished — in exchange, more or less, for a label, since the Estero was *already* managed as wilderness. By insisting on a more "pure" vision of wilderness at Point Reyes than was ever imagined by the legislators who made the designation, or by the authors of the original 1964 Wilderness Act, environmental advocates have become willing to sacrifice the relative wild for an idealized one.

\* \* \*

Since DBOC's closure, I have been thinking about ephemerality and its role in conservation work; sometimes the best response might be to simply let something go rather than try to "save" it. Yet while it might be good to embrace ephemerality now and then, one should scrutinize the circumstances when a resource or a land use is being forced out. There is a crucial difference between disappearing — there's an element of self-willed-ness in that, isn't there? (which is the fundamental meaning of *wild*) — and being disappeared.

What drew me to my particular kind of work, environmental history, was in large part my experiences as a child at a biological field station where my father did summer research in Colorado, surrounded by high-altitude wilderness yet finding within it traces of its silver mining past: a tumbled-down cabin here, the mouth of a mineshaft there, an old iron cookstove rusting unexpectedly in a field of lupine and columbine. Mining was not forced from these slopes except by economic forces; when the price of silver was high, people crowded into this valley and tolerated brutal winters for the lure of riches, but when it dropped, they too dropped away. I would not want to see those remaining traces in the landscape "preserved" or restored—letting them gradually fade seems like the most respectful response, allowing the stories to persist in the landscape without propping them up. But I similarly would not want to see them removed and all traces expunged, a purifying of the landscape to create an artificially and unnecessarily pristine wilderness.

And that is what I see under way at Point Reyes: that kind of forceful erasure, not of historic artifacts in this case, but of present-day land uses and present-day families—a community with a long history of connection to place being valued less than a label of "first marine wilderness on the West Coast." Drakes Estero was *already* wild, and removing the oyster farm does not improve its wildness. If anything, such an erased landscape will be more controlled, more constructed, and far less self-willed of a place. Those who are fundamentalist about wilderness, demanding a purist definition of these landscapes, are driving humans and nature farther apart; what we need is a return to a more pragmatic vision of wilderness, one that recognizes that not all human actions diminish wildness, so as to make room for the relative wild once more.

## Notes

1. Karen Taylor-Goodrich, associate director for Visitor and Resource Protection, National Park Service, testifying before the House Subcommittee on National Parks, Forests, and Public Lands, regarding H.R. 3022, a bill to designate lands in Sequoia-Kings Canyon as wilderness, October 30, 2007.

2. US Senate, hearings dated March 2, 1976, at 355–57. Many of the organizations that Mr. Friedman represented also sent in letters of support, all explicitly endorsing the Citizen Advisory Commission's recommendations. See, for instance, a letter from the Marin Conservation League president, Robert F. Raab, US Senate, hearings dated March 2, 1976, at 369.

3. US Congress, House Hearings (September 1976), at 4–6.

4. US Congress, House Hearings (September 1976), pages 2–3 of Curry's written statement (between pp. 17 and 18 in the document). Curry testified, "Our report was concerned and did not include the tidelands because the State retained jurisdiction over those mineral rights, but I am sure that we would have no objection to those being designated as 'potential wilderness' additions if the committee chose to act. I think that is consistent with that element in our report."

5. Letter from Frederick Smith, executive director, Environmental Action Committee of West Marin, published in the *Point Reyes Light*, January 21, 2009.

## References

Ames, Michael. "The Oyster Shell Game." *Newsweek*, January 18, 2015.

Cronon, William. "The Trouble with Wilderness: A Response." *Environmental History* 1, no. 1 (1996): 47–55.

———. "The Trouble with Wilderness; or, Getting Back to the Wrong Nature." In *Uncommon Ground: Toward Reinventing Nature*, edited by William Cronon, 69–90. New York: W. W. Norton, 1995.

DeRooy, Carola, and Dewey Livingston. *Point Reyes Peninsula: Olema, Point Reyes Station, and Inverness*. Charleston, SC: Arcadia, 2008.

Harvey, Mark. *Wilderness Forever: Howard Zahniser and the Path to the Wilderness Act*. Seattle: University of Washington Press, 2005.

Kovner, Guy. "Cleanup Transforms Drakes Estero." *Santa Rosa Press Democrat*, March 23, 2015.

Marin County Board of Supervisors. "Regular Meeting of the Marin County Board of Supervisors Held Tuesday, May 8, 2007, at 9:12 a.m." http://www.marincounty.org/depts/bs/meeting-archive.

Meine, Curt. *Aldo Leopold: His Life and Work*. Madison: University of Wisconsin Press, 1988.

National Park Service. "Drakes Estero: A Sheltered Wilderness Estuary." Originally published September 2006, revised and reissued a number of times (April 1, 2007; May 8, 2007; May 11, 2007; and a version not released publicly, dated July 27, 2007).

National Research Council. *Shellfish Mariculture in Drakes Estero, Point Reyes National Seashore, California*. Washington, DC: National Academies Press, 2009.

*Point Reyes Light*. "Wilderness Bill Gets Boost." September 16, 1976.

Pollan, Michael. Letter to Senator Dianne Feinstein, dated October 5, 2012.

Sundergill, Ron, and Sara Barth. "Why Oyster Company Must Go." *Marin Independent Journal*, November 6, 2007.

Sutter, Paul. *Driven Wild: How the Fight against Automobiles Launched the Modern Wilderness Movement.* Seattle: University of Washington Press, 2002.

Turner, James Morton. "The Politics of Modern Wilderness." In *American Wilderness: A New History*, edited by Michael Lewis, 243–61. New York: Oxford University Press, 2007.

———. *The Promise of Wilderness: American Environmental Politics since 1964.* Seattle: University of Washington Press, 2012.

Watt, Laura A. "Oyster Farm Was Intended to Stay." *Marin Independent Journal*, November 18, 2007.

———. *The Paradox of Preservation: Wilderness and Working Landscapes at Point Reyes National Seashore.* Berkeley: University of California Press, 2016.

———. "The Trouble with Preservation, or, Getting Back to the Wrong Term for Wilderness Protection: A Case Study at Point Reyes National Seashore." *Yearbook of the Association of Pacific Coast Geographers* 64 (2002): 55–72.

Wilderness Act of 1964 (Pub. L. No. 88-577). 16 U.S.C. 1131–36.

# 12

## INHABITING THE ALASKAN WILD

*Margot Higgins*

On tattered calendars pressed between jars of canned salmon, moose salami, cranberry relish, and pickled garlic, Mark Vail, a long-term resident of Wrangell–St. Elias National Park and Preserve, has been marking daily natural history observations. Nearly all of these records stem from his direct experience on the property he purchased in 1983. Faded entries recall his initial grizzly bear, lynx, and wolf sightings, as well as the first jars of cranberry liquor, cauliflower pickles, and spruce beer that he preserved. For a ten-year period, he only left the park two times—and barely so, at that. Through over thirty years of concentrated time living on private land inside the park, Mark has gained a unique perspective on the changes that have occurred in his backyard.

In late August, I arrive at Mark's mailbox to find a hand-drawn map with illustrations that direct visitors along a maze of trails. He greets me and we walk together, passing the garden beds, arranged beautifully in a pattern of disk flowers, which Mark famously fertilizes with moose droppings. Passing an enormous pile of black bear scat, he points to the red clumps within it and explains that the digestive juice of bears helps germinate these rose hip seeds. Mark knows the thrushes have been picking through the maggot hatch that typically follows the late summer decay of mushrooms on his property. With thoughts of flavoring winter stews, he selects and dries a handful of these mushrooms, which will be combined with the vegetables from his gardens and perhaps a little moose meat. Mark's subsistence livelihood depends on knowledge of the environment as a whole. It is not enough to understand the behavior of swans, moose, lynx, or bears. He needs to understand what they eat and how they behave during a big snow year or after a forest fire.

Soon the summer tourists and seasonal residents will depart and Mark will be one of fewer than forty people who reside permanently in the McCarthy community ten miles from his home. Though he is careful not to advertise his favorite gathering spots, he is willing to share the array of recipes he has invented. He donates the extra produce from his gardens to local families as well as students and writers, including me, who travel from afar to glimpse this lifestyle on his home ground. For many years, Mark has offered free classes on canning, preserving, and pickling. From his lessons, children and adults have learned how to spin sled dog fur or wild goat hair into intricate wall tapestries. I want to draw out this thread and spin more stories that capture this increasingly rare way of life.

Through summers spent guiding college students in the trail-less, backcountry, "wilderness" terrain of Wrangell–St. Elias National Park, I found a connection to the steadfast current of life forces that were both beyond and in congruence with my own. We crossed gray, silty, undammed, braided rivers that flowed through wide, U-shaped valleys and marveled at the profusion of blueberries until we were literally blue in the face. Carrying field guides and journals, we learned how to distinguish the upward trill of the Swainson's thrush from the downward song of a hermit thrush. We traveled across glaciers and felt the pulse of energy that radiates from each step along the deep-blue crevices formed by crevasses, streams, and moraines. We witnessed firsthand the age-old transformation of glacial deposition to dryas and willow, suspended seeds coming to life. All of this reinforced the notions of wilderness as a static landscape that is largely devoid of people.

Yet time spent observing the shifting dynamic of ancient ice also gave the impression of nature as unstable. I wondered: Could wilderness be frozen by static legislation? On one of my first visits to Alaska, I experienced the intense wake-up call of falling on the concrete of compressed ice that is thousands of years old, which had appeared, misleadingly, to be as soft as spring corn snow. Startled, I realized that I wanted to dig more deeply into my relationship with this landscape of dynamic change, to see what was solid and what only appeared so. I was compelled to pursue a doctoral degree that would help me better understand this dynamic and that would give me better tools to teach others. I wanted to know how people adapt to place and how wilderness might not merely be a matter of legislation but might also be manifested in the actions and intimate observations of people like Mark. I began to learn that such human lifeways are nested within changes that lie beyond local control, shaped by environmental and political conditions.

I set out to gather accounts of observations from park residents who have an unusual relationship with our national park system. More than five thousand people in and around Wrangell–St. Elias have been guaranteed the right of protecting their traditional livelihoods through the unusual enabling legislation of this park. Many of these territorial claims and hunting-gathering uses predate the arrival of the National Park Service (NPS) in 1980. It was then that the NPS began the trials of managing what historian Theodore Catton (1997) called "inhabited wilderness."

Inhabited wilderness is the result of unusual compromise, an effort to honor traditional livelihoods while protecting land within the framework of national wilderness policy. In 1980, the Alaska National Interest Lands Conservation Act (ANILCA) designated ten national parks and fifty-six million acres of legislatively protected wilderness—tripling the amount of wilderness in the United States and doubling the size of the national park system. But while the scope of the act was noteworthy, its contents marked an even more significant chapter in American wilderness policy. Following years of contentious negotiation in Congress, the act mandated the continuation of traditional lifestyles by both "Native and non-native rural residents."

Passed by Congress and the single stroke of President Carter's pen, ANILCA angered rural Alaskans, who viewed the legislation as an imposition from faraway Washington, DC. The act triggered immense protest from those who chafed at the idea of being under increased governance by bureaucrats who sat comfortably behind desks rather than engaging in the outdoors. From the time it was introduced in the US House of Representatives in 1977 until it was enacted in 1980, ANILCA was considered in more than a dozen versions. This contentious process generated thousands of pages of documentation that revealed a wide range of American attitudes toward wilderness and the role that humans play in ecological systems.

Wilderness preservation in Alaska was largely promoted by the same advocates who had campaigned for the establishment of the Wilderness Act in 1964. Hailing mostly from the lower forty-eight states, many of them had never set foot in Alaska. In order to gain support from rural Alaskans, they reached out to Native Alaskan people with the promise that establishing wilderness would also guarantee the continuation of the livelihoods they had evolved with for thousands of years. It was pushback from nonnative Alaskans that established the final compromise of allowing both Native and nonnative Alaskans to continue living in national parks (Turner 2012). Permitting both communities to continue living in the wilderness was an experiment driven by conflicting politics. Today there are twenty-three resident-zone communities in and around the National Park and Preserve, where rural Alaska residents are eligible for subsistence usage of parklands, including hunting, fishing, trapping, wood gathering, and small-scale mining—a use of resources that the wilderness advocacy community strongly opposed. Through my research, I wanted to explore these constraints and possibilities.

Of course, the adaptive experience of living with dynamic nonhuman systems dates back thousands of years. There are four distinct Alaska Native groups and a mix of Alaska Native people who have specific tribal affiliations with deep ties to the lands of Wrangell–St. Elias National Park and Preserve. Historically, the Ahtna and Upper Tanana Athabascans resided in the interior of the park before European settlers encroached on their land in the late 1700s.

There is no word for *wilderness* in any Native Alaskan language. Native Alaskan articulations of the wilderness must grapple with the patent absurdity of placing a geographical boundary around nature. Many Native Alaskans believe that legislatively mandated wilderness designation is yet another wave of the colonial theft that has continuously displaced Native Americans in the name of "higher" use, from manifest destiny to national parks. They claim that ANILCA has further restricted their access to traditional hunting and fishing grounds. Despite the fact that they were promised employment as a part of park designation, Native Alaskans have few jobs to show for it.

Life in rural Alaska is an experience I could not properly imagine until I spent several seasons there. Removed from the Internet, my days were occupied by long moments of silence and meaningful human engagement. It is still customary for people to stop by without calling or texting. They often carry something to eat, a musical instrument, or an urgent request for help. Floods blow out footbridges, and heavy rain pours through the walls. I encountered moose and bears on long bike rides in between interviews. I hauled my water a quarter of a mile from the stream to a cabin that had been constructed decades prior with hand-peeled logs. I attempted, sometimes unsuccessfully, to power my voice recorder with a small roll-up solar panel.

One quiet evening, alone, while weeding in the garden, I locked eyes with a lynx for what seemed like an hour.

It is necessary here to adjust your life to the intense cycles of the midnight sun and the drawn-out dark days of winter. People treasure the taste of black bears and grouse fed by berries. They long for the salty flavor of herring eggs laid on seaweed, raw sockeye salmon straight from the Copper River, or even salmon eyeballs. Rusty trucks, copper barrels, and Blazo cans are a common site in the Wrangells, as are bear-scratched, weather-scabbed cabins built with hand-peeled logs and roofs covered with lichen and moss as insulation. Residents seem to be continually in a state of fixing busted windows because of bears and avalanches or patching holes that voles have incised in their walls. They axe multiple cords of wood to heat them through long seasons of cold. (It is not unusual for temperatures to drop well below zero, though that seems to be happening less frequently these days as plant ranges shift, ice breakup arrives earlier, and heavy winter parkas are pulled out later in the fall than in the past.)

The stories that the non-Native residents of Wrangell–St. Elias shared with me often recalled a recent history that seems familiar; they expressed romantic nostalgia for the lifestyle their great-grandparents and even their parents might have lived. And yet, for many Americans, such attachments require no small amount of interpretation.

A nonresident of the park might ask, as I have, "How have you managed to stick it out in this remote place? And why?" The first of those questions, the how, initiates a long but vivid conversation about such subjects as severe weather, hard times, ingenuity, adaptability, concern for the land, government policies (the despised ones and the depended-upon ones), distance from global markets, and the healthy aspects of growing, gathering, and hunting your own food.

The second question—why?—is not so neatly answered. Some people cite close-knit communities and deep long-term relationships that are like family. One man said, "It is a place where we are limited by our environment and we learn to live with fewer comforts." From the home he designed "thriftily" and constructed for its "functionality," he explained his decision to live in an area with less connectivity to the wasteful patterns of modern society.

A professor who is a resident in the area told me, "I like living in a community where the teenage boy in the leather jacket also has an intimate knowledge of birdsongs." For other residents, living in the park is about confronting the scale of our human lives. "It is possible in this place to be right in the face of huge physical biological geological change happening in multiple contexts at once," a woman said. Another commented, "This place is high energy. It feeds off the glacier that literally spills into this community. Many people like myself are drawn to that. Some go crazy." A man who has lived in the area since the 1970s described a deliberate spiritual path intricately connected to the patterns of ice and water in the winter season. "We created a secular monastery," he said, "a place of spiritual quest and intent. We wanted to live with the snow and ice as it comes, September through April, especially when nothing else was going on. For us it was just snow and ice. We lived with the slight change in the wind, humidity, crystal in snow. These marked the changes from one day to the next."

But for some of the park's most tenacious residents, the only articulable answer is seemingly circular—a reference back to the place itself. At eighty-two years old, the longest resident of one park community repeatedly told me when I was a guest on his property for a summer, "I like getting up and looking at what I get to look at every morning." He referred several times to his experience of spotting black and grizzly bears, seasonal songbirds, a moose and its calves—and a lynx here and there—though the large kitchen window.

In recent years, there has been a substantial out-migration of the permanent population from most park communities. People I have talked to explain the exodus by pointing to poor fishing seasons, the rising cost of food and fuel, and reduced state spending on schools, roads, and other maintenance. Despite this pattern, however, tourism has been on the rise, as have the numbers of seasonal homeowners and investors. Wrangell–St. Elias is increasingly an "emerging rural landscape," as wilderness scholar Nathan Sayre notes, where the goals of capital investment are not limited to the fur trade and mining natural resource commodities of the past. The area now attracts investments for tourism, housing development, and continued environmental conservation, as well as the "speculation in all of these" (437).

Inhabited wilderness is becoming less inhabited, at least in this case—a situation that is further exacerbated by the growing urbanized tourist economy, exposing a difference in land management and use that deeply divides the local population from NPS staff who are beholden to a different set of objectives for the park. Tourists generally come to Wrangell–St. Elias to experience the largest wilderness area in the US national park system. Like me, they are often surprised by its continued history of inhabitation.

One key point of contention regarding how the area has been used—and will be used—dates to the turn of the twentieth century, when white men discovered dense deposits of copper ore in the middle of the Wrangells. The prospects were so promising that, despite the significant obstacles created by hard weather and shifting glaciers, wealthy New York investors built a twenty-million-dollar railroad to transport copper from the interior of the Wrangells to the coast in Cordova. Painstakingly constructed, the railway required 129 bridges, and when the Kennecott[1] mines closed in 1938, rain, snow, and melting glacier water quickly demolished many of these impressive yet fragile structures.

One such bridge—at the end of the railroad line in Kennicott—lasted far longer than was initially anticipated. Following a heavy rainy season in the fall of 2006, flash floods rushed through the former company town, which had become a premier tourist attraction. The railway trestle that had functioned as a pedestrian bridge over National Creek finally collapsed. In order to prevent future damage of the two-million-dollar replacement bridge, the NPS proposed creating a creek diversion with treated lumbers imported from Washington State. This proposal quickly attracted the attention of locals: they were concerned about the potential for the chemically treated wood to leak toxins into their primary source of drinking water. Tensions were heightened by agency responses, which were largely dictated by historical and distant policy making in regional and DC offices.

The contentious meetings that followed were punctuated by strong assertions of national and local power. Many residents of this community argued that the NPS was "Disney-fying" the area and recreating a company town that, like its predecessor, had destructive impacts on people and the land. Pressured by local residents, the NPS replaced the proposal for using treated timbers to stabilize the creek with an alternative plan. In order to soften the velocity of a potential flood and to prevent damaging the new bridge, NPS crews placed large slate-gray boulders in a uniform sequence below the creek. Most of the material needs for this construction would have been readily available locally. Nevertheless, because NPS policy prohibits the agency from mining anything on national park land, rocks were trucked in from two hundred miles away in Valdez.

One local reflected, "There's now a line of boulders in National Creek in Kennicott from coastal Valdez. The image I have is this little parade—this line of trucks carrying boulders this way." Pointing to the glacier that runs along the side of the Kennicott community, he asked, "What is out here carrying rocks the other way? There are enough rocks moving this way to probably provide all of the material needs of the western United States."

"Part of the attraction of being here is that Mother Nature is calling the shots and we weave our lives around that," said an inholder who has witnessed the area's development since the early 1990s. She commented, "The rocks below National Creek look human placed. I know they are not local rocks. The more we fill the landscape with false truths, the more difficult it is to read it, the less we trust it, and the less inclined we might be to bother about protecting it."

While this person acknowledges that there are many visible human-created geologic stories on the landscape from the mining era, she believes that the NPS should be more careful about the new story it is imprinting in the area, especially with regard to how, in this case, the geologic imprint impacts local stories and access to resources such as clean drinking water. "It is a big fat lie in the middle of an authentic story."

Within a small community of in-holders that lies parallel to a melting glacier, the NPS celebrates the height of our industrial mining era. Melting glacier water revealed to me the conflict between the physical reality of a dynamic landscape and attempts to create stability within it. In the Wrangells, it is challenging for the NPS to solidify its own reputation as a federal agency that is often less attached and responsible to local concerns than it is to adhering to national policies. The gap of NPS engagement with locals, however, is in part what prevents it from implementing more cost-effective, well-accepted solutions to problems. The resulting dismissive narratives about NPS decisions that are told by residents can create barriers to collaboration and innovation.

Mark is one resident who cautiously welcomed park designation, and he continues to engage with the NPS around management decisions, encouraging others to do so as well. "How do you protect something? How do you idolize that? How do you express that even to yourselves, to the community?" he asks. "It's a hard dilemma, and communication is the only way to do it. That's why we're constantly haranguing the park service to communicate, communicate, communicate. I have been haranguing my community because a few weeks ago the park service got fourteen people to tell us what they were going to do this summer and only three people from the community came to listen."

In more recent years, interpretations of the complicated ANILCA legislation continue to play out. Of course, it is largely through ANILCA that such cherished traditional practices and livelihoods are able to continue. A wild way of living is vanishing, yet it is a right that only a few have benefitted from. Coevolving with the landscape is a distinct privilege that has evolved—not only because of the ANILCA legislation but also due to the prior ravages of disease, colonialism, and the direct displacement of Native people.

Given this problematic history, throughout the United States there are impassioned debates about wilderness designation and whether or not citizens should support more models of wilderness that include people and their livelihoods. Wrangell–St. Elias is a place where native people and nonnative people, native rocks and nonnative rocks, local governance and federal governance rub up against one another and sometimes collide. Some people visit; some people live it.

Alaska has one of the fastest-growing populations in the United States. Some predict this trend will continue as more people flock to the north because of a warming climate. The NPS is tasked with monitoring changes in the land. But at present there is only one full-time ecologist and one full-time biologist in the park. People who are attuned to the direction of the wind, when the maggots hatch, river bank erosion, and the degree to which salmon runs are shifting and declining might have something to offer in documenting such changes in the land and water. If we are lucky enough to visit Wrangell–St. Elias, those of us from urban California, who typically hike through the wilderness in Gortex jackets with field guides and binoculars, might have something to learn about attentiveness as a daily practice, as a way of cultivating more compassion for protecting what is wild, in rural or urban settings.

I take some of these lessons back home to the mild manicured landscape of UC Berkeley. On the first day of the semester, I ask my students, "What direction does your bedroom window face? Where does your water and heat come from? Can you name five native plant and bird species?" Although they may have been previously engaged in environmental causes, these students are stunned when they fail this first quiz. Throughout the course of the semester, I try to teach them to slow down, to record their observations in a journal, to understand where their food comes from. I want them to realize that "wilderness" goes beyond lines on a map and that even in urban California it can inspire a way of living that is accessible, responsive, and adapted to place.

### Note

1. The Kennecott Company name is spelled with an *e*, while the name of the town, glacier, and river is spelled with an *i*. Kennicott is also the frequent spelling used among residents of that community.

## References

Sayre, Nathan F. "Commentary: Scale, Rent, and Symbolic Capital: Political Economy and Emerging Rural Landscapes." *GeoJournal* 76 (2011): 437–39.

Turner, James Morton. "Alaska: 'The Last Chance to Do It Right the First Time.'" In *The Promise of Wilderness: American Environmental Politics since 1964*, 141–81. Seattle: University of Washington Press, 2012.

# 13

## WILDERNESS IN FOUR PARTS, OR WHY WE CANNOT MENTION MY GREAT-GRANDFATHER'S NAME

*Aaron Abeyta*

### I

The snow came early that year, 1949, in what people now call the Toltec Wilderness, and it seems a miracle—by modern sentiments, anyway—that my father, then a twelve-year-old boy, would be allowed to ride the seventeen miles, through a blizzard, so that he and my *abuelito* might rescue the herd.

There is not any particular magic to the way a sheep, or a herd of sheep, die in a blizzard. It is not from the cold. They huddle together for warmth, their bodies like pills in a bottle, and the snow falls, and the snow falls, and the snow falls, and the sheep do not move, and the sheep do not move, and in the most basic sense of the word, they suffocate beneath a slow avalanche of accumulated snow. Perhaps it is akin to insanity that even when the snow is at their eyes, the sheep remain as motionless as windows, witness to their own demise but patient—a primal faith that their shepherd will arrive and save them.

Even at twelve years old, you become aware of the way things die and, reciprocally, how things survive. A person takes note, especially, of the death around them. If you live on a ranch for any period of time, you come to know the many ways that things can die; a calf dies near a pile of concrete at a fence line in the middle of a June prairie, bitten by a rattlesnake, and even when you are a grown man, you will still hear that mama cow and her cries into a bright and cloudless day. Loss has such an echo, eternal.

Or maybe you will inherit a racehorse from a family friend, and you will marvel at the height and elegance of the animal but struggle to understand why, come winter, despite the snow everywhere, the animal refuses to eat the same hay as the other animals. You will marvel at its inability to survive. Named Master Bars, he was the most beautiful animal any of us had ever seen, but he died on a January night, and he was not beautiful in death: a frozen emaciated corpse, eighteen ribs all touchable and arching into the January day. It will be his ribs that are welded to the frozen new light of dawn and into your memory.

You will not remember every lamb that was abandoned or sat upon by its mother, but you will place them all in a collective memory and you will look back at times to their collective death, and it will become clear why the lambs that survive jump and play in the March mornings; they jump and play so that those watching might be healed. Like the lambs, you will make a singular memory for all the calves born into snowbanks, their bodies frozen within minutes, bodies so motionless you might swear they were made of wood or of ice, bodies that tried to live but could not. There will be Suffolk rams bloated and green at the edge of alfalfa fields; a saddle horse mistaken for a bear; a semi-trailer that did not slow down as the herd of sheep crossed the road to water; a stock tank vandalized by environmentalists, thirty-five thousand gallons turning the earth around the trough to a mud so deep and thick that seven cows and one calf would die there — but not before the crows had landed on their struggling bodies and eaten out the still-living eyes of the trapped cattle — and you will remember how your father fell to his knees in the very same mud and gave the calf CPR, and you will feel relief and shame for feeling relief when he finally gives up trying to save the animal. You will come to know, eventually, all the different ways an animal can die, and sometimes you might even think it beautiful and that just makes you human, the way we meet fear and loss and name it something else.

So in the regular world, it would not make sense to send a twelve-year-old, on horseback, into a blizzard for a seventeen-mile ride to the high country, the place where the storm is from and has always lived — but to all of us who know death so well, it makes a perfect sort of sense. It is our job—thankless at times, but necessary—to keep death at bay; this is why my father and his father saddled up in the 4 a.m. darkness of a late September, the moon a vapid pearl behind the storm clouds, and rode off toward the west and toward elevation, and toward a herd of sheep that would all suffocate to death if they did not do so.

## MY GREAT-GRANDFATHER'S NAME

Four children and a wife were left behind when my *bisabuelo* abandoned his family—the oldest, an eight-year-old, was my *abuelito*, Amos. He tells of how cold the house was. He recounts the trains that passed on their journey north, bellies of coal brimming black as burnt-out suns, the trucks of the gondolas rocking along the standard gauge line, the slow and heavy clatter of a train as it passes, a noise that recedes in a drawn out minor key. There, the boy who would become my grandpa hunted for fallen coal in the gray ballast that was the same color of gray as the track bed that was the same gray color of the day as it passed. I imagine how each piece of found coal was equal parts joy and resentment. The man I knew, so many years later, would not have come home until the bag was full, and I imagine that black, bituminous load, there upon his shoulder, was a burden that even time did not, or could not, lift. The man I knew as my grandfather was made of stone, and my awe of him was only tempered by a greater fear. There is some speculation that, perhaps, I have inherited his shoulders.

My great-grandfather was named Serafin; it was also my *abuelito*'s middle name. My great-grandfather's name was not a sin, but it was treated as such. Yet fragments of words and stories have fallen to the floor and been collected in a dustpan of partial stories, memories, and impossible clues that were set to flame by anger and antipathy. I know this. He abandoned his family, an abandonment so complete that my grandfather chose erasure, an absolute and collective deletion of his father, as his vengeance.

## *II*

**the elk calf**
i am looking for scattered sheep in the wilderness
the herder has fallen ill i am on foot the horse
is in the wind the horse is smoke the horse is pollen
the horse is ghost and the dogs have no loyalty to me

i am walking the meadows of rincon bonito
the old men call the spruce at the meadow's edge
los brazos translated the name means arms but
the ancient meaning is shadow and silence

i must enter the spruce my abuelito's
voice tells me i must get the count we must
know how many have died how many will not
return to the llanos south of home we must
know how much of our winter work
has been lost here in this late june

i will not find every sheep
it has been too long the herder
sick for five days i am only eleven
but i know what death is i have
seen the violence of what
dogs can do the neck wounds that
only coyotes make

i imagine the calf female
weigh her with my eyes
forty pounds i tell myself
the clearing is small no grass
small bits of bark twigs dark as morning dark
spruce needles the gold of dying things
cling to the still wet animal her amniotic sac
a yellow shawl on her shoulders ears wet
the placenta and cord at her nose
i pray to God silently
that i am allowed to witness this
pray that the cow elk
is only at the spruce edge of the forest
her large and sleek body somehow brought
into the safety of a shadow
human eyes cannot penetrate
i pray because that is what my abuelita
has taught me to do
pray that my being here this accident
will not mean the death of this animal

i dare not touch what
my touch will doom for having touched

as a man i carry this anger it is
untraceable yet i know my father taught it to me
with his blood with his stories he loved
all of us enough to teach us not to trust
even so his eyes have in them the dark well of mercy
this vine of flower is watered by fire and it is my life

beyond the newborn elk calf
the spruce drop down a slight slope
light enters in razors of dust pillars of gold

at the edge of the clearing there are
six sheep buried in the duff their bodies
bloated bellies green and blue
necks broken

i am eleven the horse is
in the wind and the dogs
have no loyalty there are
two ravens at the edge
of the trees the invisible
magpies are crying into the day

i look back toward the elk calf
i do not know what to do i
am alone i pray because that
is what i have been taught to do

i pray for myself i pray
for the count and perhaps i
pray that too much death will
not enter into my life *i must
have prayed for something like that*

o dear and brutal day
do not seep into my young heart

dear Lord and dear Saint Francis
look over the newborn elk calf
may her mother hear her chirp
may her mother lift her head and
run toward the sound and may
all living things that have not
yet done so dear Lord
may they suckle

o dear and brutal day
whose light is pillars through dark
arms of spruce may the horse
return to camp and may the dogs
always be loyal

o dear and brutal day
here where i stand at the edge of
death and birth protect me

o small voice that was me
a thousand years ago tell me
which way the bear has gone and
lead me away toward safety
and living sheep small voice
that was me so long ago let me
sing later let me not know too
much anger let me sing forgiveness

remind me o small voice
that my father has sent me here alone
because he loves me and understands
that men must know their fear
if they are ever to love

o dear and brutal day
heal the herder and lead
the horse home lead too
the mother elk to her calf
lead her to lick the newborn clean
lead her to eat placenta and cord

lead her to swallow the danger
the scent of these things brings

my abuelita has taught me to pray
she tells me our faith is made of
three pillars prayer penance
and action that there are eight
types of literature in the bible
this is one of them she has
taught me to pray

i would give away most anything
to hear her voice again i would
give away words and anger
i would give away fear and joy
i would give away this abyss between
life and death i would
give away this spruce and
every wilderness to
have her lead me in prayer
just one more time

i am just a boy
she died the winter before

i ask her to ask God to
save the elk calf i did
not dare touch i ask her
to walk me back to the
open meadow and i ask
that the count not grow
too high or too heavy
for my young body to bear

*III*

The meadow is five miles long and at its narrowest southern point turns, with the creek, to the west where it funnels toward a stand of spruce and pine, a place the herders called Rivera's; the Rivera cabin rests at the narrow confluence of trees, creek, and meadow. When the days were long and the moon was good enough to see by, the herders would go there for stories of bears, coyotes, hand-caught brook trout, and elk turned to the golden dark of summer. There was wine and whiskey. It was a given that the horses knew the way back to camp.

When the snow began that September, the herd was near Rivera's, and by the time they had reached the far end of the meadow, the snow was too deep for them to continue.

Sometime after sunup, my *abuelito* and my father reached the Los Pinos River, eleven miles from home. From the river, the horses would have to climb six miles to the northern end of the great meadow, where the herd of sheep was trapped by the still-falling snow.

The herder was a man named Fidelito, and he was a good man; he knew the snow was in a race with the flock he was paid to tend. He saddled the animals and packed the camp in a rush; nothing was covered. He pushed across the meadow, moving north toward a crest of great, round, pink and green rocks and a small stand of aspen. It was the only windbreak. The herd would not travel any farther—the snow too deep, their bodies encrusted. He worked the herd between two house-sized rocks and let them rest among the aspen where the snow was not as deep. He would wait for the snow to cease its falling or for help to arrive. There is a profound faith found, perhaps, only in sheepherders, that cannot be defined; they trust that even in the darkest moment that grace or deliverance will find them. I know this to be true. I am a living witness to such a faith; I believe and know it to exist in humans, a rare and deep human core as interminable as hope. Two feet of snow had fallen and the sky was still winter chalk, but I am positive that Fidelito knew that my grandfather had already crossed the black ribbon of water that is the Los Pinos River and that the horse would carry him to this promontory at the edge of a great meadow in the center of a greater wilderness in the midst of an even greater storm and that somehow they would prevail against the snow.

## MY GREAT-GRANDFATHER'S NAME

For a time, before he took the title from Jess Willard in 1919, Jack Dempsey went by the name Kid Blackie, and he would make his money fighting in saloons, in makeshift rings, or on warehouse platforms. The man that the world knew as the Manassa Mauler was born just a few miles from my hometown; he was, and still is, a legend. The argument goes that he had the best left hook ever, and that may be true. Even before he was champion of the world, people knew he was not to be quarreled with. To fight him during that time was a stupidity reserved for miners and prospectors, unfortunate men who did not know Dempsey or his reputation. Locals knew to steer clear of his taunts and bravado, a well-rehearsed bait to separate men from their paychecks and their senses.

Sometime after my *abuelito* passed away, I received a call from a distant relative who was working on a family tree. He had come to the portion of the tree where there was some confusion, the place that silence and a forced forgetting had carved out; he wanted to know about my grandpa, Amos.

I told him that my grandpa was a five-term sheriff of Conejos County, that he had been shot twice by a man who was made mad by *brujas*. I told him how the first shot grazed the center of my *abuelito*'s skull, just above the bridge of his nose, and the second entered below the sternum, followed the curve of a rib, and exited at the spine. I told this distant relative how, afterwards, my *abuelito*'s right leg was never right, how I tended to stare at the jagged, pale, soft-pink flower of a scar in the middle of his back as he shaved, standing at the kitchen sink while looking over the ranch he had built from hard work, anger, and resentment. I told the man on the phone that he was the toughest man I have known and will ever know.

The man on the other end of the line was not surprised. He asked me if I knew about my great-grandfather, how the rest of the family thought him to be a hero of sorts. He recounted stories his father had told to him; the admiration in his voice was audible and true. He asked me if I knew about my great-grandfather, my *bisabuelo*, fighting Dempsey in a potato cellar one late August day. I told him I did not. He did not ask if I wanted to hear the story; he simply began, presumably in the voice of his father or someone lucky enough to have been there that day:

> Saw him fight once. That's why that man hired him, because he was stronger than hell. It sure was hot that day, humid too. End of August and all the people down at the cellar had just been paid. Big, thick clouds were in the west over the mountains. Far off, lightning and thunder like it was coming toward the end of the world. I suppose it was right around evening, and most of those men had been loading hundred-pound sacks of potatoes all day. The loading docks were stacked with them, like a fortress of burlap, roots, and dark earth. Stanley Barr said he was worth two men, the way he could work. Your great-grandpa was younger then, stronger too.

He paused, the phone humming slightly. There was a great distance in that silence. I remembered it from my youth and knew it to be inherited and then taught. Then he resumed:

Everyone knew Jack. This was before he was world champ. He had left Manassa a few years before, but he would come around every now and then, visiting, I suppose. He was one of us, poor as worn-out shoes; everyone knew that. He made his money in the bars and the potato cellars, and we all knew that was why he was there that evening. He had a routine; we had all heard about it. He said he couldn't sing or nothing like that, but he would knock the tar out of any man willing to put up a few dollars. At least he was honest. Jack was meaner than hell. He could punch like a mule kicks. No one was stupid enough to take him up on his offer. We had all seen what he could do. Stanley and everyone else just laughed a bit, shook their heads, because they knew that no fool would fight Jack. That sort of stupidity was reserved for miners and prospectors. Jack went on for a few minutes and all the men, except a few, had stopped working.

Your great-grandpa was up top, ten or twelve feet off the ground, stacking hundred-pound sacks near the top of the cellar; he must have been impatient. That's the way with men that know how to work. They ain't got no patience for the ones that don't.

Everyone was just standing around taking a break as Jack was talking away the last of the light. It went silent as a funeral after your great-grandpa finally spoke up, something along the lines of "*You talk too damn much. Shut the hell up so we can get to work.*"

He wasn't a miner or a prospector, and he knew Dempsey, same as the rest of us. I guess his cup just filled up.

I suppose Jack would have fought him for free, he was so mad. Your great-grandpa came down from up top, just a few dim bulbs, a setting sun, and some far-off lightning were all we had to see by. He was dark with the dirt of all those potato fields. He had boots that were so beat down the heel was nearly gone. He must have been twenty pounds lighter than Jack.

I never saw Dempsey fight Tunney or Willard, but I saw him fight your great-grandpa. Jack walked across that platform like a storm and Serafin just stood there, his body caked with the dirt of other men's fields, his hands balled up and at his waist, like he was too tired to lift them. And I will never forget the way Jack shook his head after your great-grandpa hit him the first time, like maybe he knew he had really been hit and maybe it wasn't going to be so easy.

## IV

My father remembers this, perhaps, above all things about those two days and the sixty-five years in between: He was cold and his father laid him down beneath a spruce and then covered him with a burlap tarp; he stuffed the tarp with duff and spruce needles, and around his cold son he placed three saddles as a break against the wind, then he walked into the onyx night and into the wind that began when the snow ceased, and he shoveled and dug through the night to save the herd. In my father's retelling of the story, he does not mention sleep nor does he mention being cold. There is admiration in his voice. It is sincere, and I imagine it to be older than written words.

I wonder sometimes if those two needed that wilderness and that blizzard. I remember them in my own time, and I cannot remember them getting along, not really. My father could never please my grandpa, and I remember disliking the man for the way he treated my dad. I loved my *abuelito* too, we all did; he was what we all wanted to become. That is the way with blessings and love—their seed, no matter how fervent, must have its curse, and by curse I mean blood.

What drew my great-grandfather down from those potato sacks, and is the story true? Did he fight Jack Dempsey until both men were too tired to continue? Did he abandon his family or is the other part of the story the truer telling? Did he have an affair with the wife of his boss and did that man then frame my great-grandfather for stealing? Was his leaving a story they told the children to protect them from the fact that he was sent to jail? Can you spend your entire life forgetting the man that is your blessing and your curse?

Which is the greatest foe: the future heavyweight champion of the world; the long and arduous task of purposely forgetting; the blizzard that could have buried an entire herd alive; the blood of a father traced in the fists of his son; a wilderness of great meadows, house-sized rocks, wind, and spruce; a seventeen-mile ride through a blizzard; shoveling for twelve hours straight; walking the empty sides of train tracks looking for coal; knowing that your oldest son wishes to never speak with you again; accepting a lie as truth; murdering away reconciliation; the daily task of never being whole or satisfied? Which is the greatest foe, and how do we come to know the thing that is most like us? I ask myself these questions, and I wonder if similar questions arise in places without mountains, rivers, and trees made of shadow and silence. Surely this is the work not of nature but of human nature, the stormy and fierce heart of every human.

Dawn broke and the wind finally stopped, faith was rewarded, and my father rose from his bed of spruce needles to the sound of axe against timber, the constant drum-thump of it as two small trees were felled. The trees were tied to my grandfather's horse. Only the heads of the sheep were visible, but he rode the horse around the herd several times, clearing a path with the wake of the trees. Then he pointed the horse north, downhill, toward the river and home. Fidelito and my father urged and pushed a sheep onto the broken path, and where she previously stood, there was the smallest of openings in the snow. Another sheep entered into the opening and then she too followed. One by one, they broke free and walked after their savior toward lower ground. By noon they had crossed the river, and by nightfall they all had reached home—alive.

All my grandfathers are gone now. I still, on occasion, drive the road up from the river, and I recall the camps, the good meals, the horses that went missing, the several herders and their personalities, the animals we lost to bears or coyotes or random death. Eventually, you reach a place where the road ends and there is a snow- and wind-battered sign that reads: "Wilderness. Closed to motorized vehicles and motorized equipment." It is known as the Toltec Unit, and it is a cruel place with little water and a loneliness that is difficult to comprehend. It is the place where so many stories begin for me, their genesis traced to particular shadows, particular camps, and the way the light sometimes makes everything dance. I think of the men in my life and always they are associated with places and with stories. Sometimes the places are wild and sometimes they are on the side of a well-traveled highway. I will never know my great-grandfather, not really. I know two stories about him, one where he is a hero and one where he leaves. The stories of my *abuelito* and my father are more numerous and more complimentary. I believe that I understand them, and there is a grace in that knowing.

I reach the rocks and aspen trees where the herd was nearly buried. The natural world is intact and most likely much the same as it has been for centuries. Human presence has faded in the proper order of things. The place is made sacred by my memories and by the strength and the brief kindnesses displayed there so long ago. I look around and there are names carved into the trees, but there is no need to read them; I know all they have to say.

# PARt 3

*Urban Wild*

# 14

## WILD BLACK MARGINS

*Mistinguette Smith*

### Is This Land?

I often open my talks about black people's relationship to land by sharing a photo. It's an image of an urban street: A sidewalk of fractured concrete with chicory and tufts of grass growing in the cracks. Beyond the curb there's a blacktop road. On the other side of that road, one can glimpse a bit of neatly trimmed green lawn.

I ask, "Is this land?"

Some people are absolutely certain of their answer: This is not land. Real land is wilderness: awe-inspiring mountains, lush forest, sudden succulent blossoms in an otherwise arid desert. Others are not so sure. This might be land—perhaps if it wasn't paved or if it wasn't in the city, it could be land. Some people insist that there is land "up under" the pavement and blacktop, citing its potential in the cultivated lawn.

This might be land if there were no people here.

The notion of land-as-wilderness is an essential difference between how black people characterize their relationships to land and the predominant environmentalist view. Land-as-wilderness presumes the wild to be a place of growth without cultivation, where being is native, rooted, autochthonous. Such notions of wilderness often invoke a place of unrestrained freedom, a lost Eden of joyful relationship to land without the bonds of ownership. Today, in both popular culture and among most environmentalists I encounter, such "wilderness" is a place to which one journeys for contemplation and recreation, or a place to which one escapes through history or privilege.

This notion of the wild requires a colonial memory, one unwilling to witness long histories of indigenous naming, habitation, settlement, and cultivation. Such a notion is rare among people who are African American, Afro-Caribbean, and others whose blackness is a marker that they are neither indigenous to this land nor colonizing settlers. While blackness as a racial trope is often associated with wildness, it is also a racial identity that signifies a condition of displacement, of being uprooted from place. In America, black people have made the bricks of the built environment by hand. Black people have done the work of cultivation and domestication of wild spaces without ever being named as the ones who turned forests into plantations and pastures and who have done so under a freedom deeply constrained. And yet one would be in error to assume that blackness does not also include a joyful relationship to land that is not defined by the bonds of ownership.

So where does the wild dwell in the black imagination today? It is very much alive. It animates the ways that black people speak and think, though this is sometimes indecipherable to the untrained ear. It grows ferociously and tenuously, in sprouts that create new cracks in the sidewalk every year. It makes new uses of what our culture discards and breaks open what we think we know. The black wild is neither somewhere "out there" in a mythical agrarian southern landscape, nor is it being restored in raised-bed urban gardens on city lots.

The black wild is interstitial, the thing that's thriving in the in-between. In nature, an ecotone is the place where two ecologies—for instance, field and forest—meet and are in tension. The black wild is the rich ecotone where built and natural ecologies meet and are in tension, creating places where diverse forms of life compete and cooperate. These wild black margins are the places toward which I have turned my listening ear.

## *Black/Land*

I did not understand how marginal black perspectives on land were until my work began to take me to conservation conferences and environmental leadership retreats. In those rooms, gentleman farmers, environmental scientists, and conservation folk spoke with awe and admiration of this thing called Land. It was their religion, complete with its own mysterious texts by Aldo Leopold and Wendell Berry. I did not know these apostles, so I listened to them closely, with a kind of anthropological curiosity. They described rolling fields of pasture as if they were feats of nature, not the result of can't-see-to-can't-see backbreak of human beings using dangerous tools. They made in-group jokes about Muir and his wilderness as primeval home. Unlike Muir, the wilderness these men talked about—for these talkers were mostly men—never had tapeworms or malaria or things that made them itch.

In rooms with agricultural leanings, I heard about Rudolph Steiner's theories of biodynamic farming without ever a mention of an agronomist named George Washington Carver. In all the discussions of U-pick orchards and urban Community Supported Agriculture, not once did I hear cited the name of Booker T. Whately.[1] So many black ideas sprang full-blown from the mouths of these nonblack speakers without the courtesy of attribution. I wondered if people with eyes and skin the color of earth became invisible when standing on it, studying it, and tending it with every kind of care.

As a result of hearing these conversations, I began asking black people to tell me their stories about race and their relationships to land. I expected to hear stories of alienation: some of the elders I interviewed are survivors of Southern debt-bondage sharecropping or land loss through the displacement of urban renewal. To my surprise, I often heard stories filled with images of unbounded intimacy with the natural world:

> I knew as a five-year-old can know things. I knew that sunlight made my skin because my skin looked just like those hills in the distance and looked just like the land around me, like a burnt-honey brown. And that sky flowed in my veins. So I was made by earth, made in earth, made by it, and I believed it. And I heard words like "colored," but within my family, among my parents, I never heard conversations about race. I didn't know it. It wasn't part of anything that I could know. So when I heard the word "colored," I thought that meant the coloring by the light of the sun, by the blue of the sky, by the land itself.

That five-year-old, who grew up in urban San Francisco and Los Angeles, quickly came to understand the relationship between race and being displaced from the land of which she was a part. As an adult, she is a geologist who feels most free and at home in her skin when she is in the external wild:

> I live deliberately in a rural area where my home is on two acres. And I have a large garden, and I can walk out from my house onto a dirt road. I cross a stone wall and I'm in open space. That walk could take me for hours and hours before I see anyone else. . . . And there's a rocky hill that's right behind my house called Long Hill, and I make a point of walking up that ledge and being there. It's a ledge, a big knob of schist covered by hemlocks and some deciduous trees. It's a place to mark the seasons, the passage of the seasons, the passage of light and the sea weathering. The weathering of the hill, erosion, it occurs at such a slow pace. I just hope that the weather and erosion that I'm feeling, that I'm experiencing as I get older, is also as gentle and as graceful as it is on that hill.

For a child to understand the brownness of her body as something made by the natural landscape rather than by an ideology of unnatural hatred reveals an internal wilderness—a sense of freedom and rootedness—that is more fundamental than the meaning imputed to her skin color by the larger culture. Her contemplative relationship to the geology and topography of her home is related to her family history of moving from place to place to escape the bonds of racism. Instead of bitterness, that history propels her to find an internal and external uncultivated place, which becomes "gentle" and "graceful" with the passage of time.

Black people do not need to live in rural or woodland places to find the wild as an internal and external synchrony. As I listen to black land stories, I am struck by how often the experience of wilderness is described as a liminal space where one thing meets and becomes part of another: where the sea meets the land or the place where a river overflows its banks. When black people tell me about their relationships to land, their experience of the undomesticated natural world is not distant or distinct from cities. It is often embedded in the built environment, creating soft boundaries in highly urbanized places. A longtime resident of Detroit described this in detail:

> I do not thrive in an urban setting. I must have nature around me to feel comfortable and whole. And, in some ways, Detroit is a cool city for that reason 'cause it's pretty sparsely populated! (*laughter*) It doesn't always feel like a dense Manhattan. There are places that feel cool and dense, but there are fields where there are pheasants and crickets and stuff like that. You don't

have to go far in Detroit to get that.... I'm always seeking the natural, as many natural spaces as I can, no matter where I am. Away from maddening crowds. I am a nature seeker. Sometimes I feel like it's true that many African Americans, especially since I've been in Detroit, they don't know the natural, they don't really know or appreciate the natural spaces, and there's a fear of it.... I don't have a lot of friends that enjoy [my] kind of interaction with the natural environment.

As someone who enjoys hiking and camping, this woman also finds the wild in her own backyard, as open space in a depopulated city reverts to native grassland habitat for feral pheasants. As a nature seeker, her relationship to land is defined not by economic ownership but rather as access to an "emotional space" suitable for acts of spiritual and cultural pilgrimage: "The Grand Canyon was one of the most transformative experiences of my life. I've been there twice. My family still laughs because I [arrived and] just started bursting into tears. I mean, it's just the magnificence of it. Again, it's like the mountain. I like that mountain, that foreverness." Unlike narratives of the wild as uninhabited, ahistoric wilderness, access to such transformative spaces is a palimpsest, where a present-day camping experience is overwritten on a collective past that bleeds through:

> I feel like that big boat trip made us afraid of water, and that little stint in the South made us afraid of woods, and it took away, it took away our humanity...
>
> Funny story: We were in the woods on the Fourth of July weekend once... near a city that was doing fireworks. Our campsite was really not far. So we were camping and fireworks go off, and I said, "Oh my gosh, I feel so patriotic! Let's sing the national anthem!" And my kids, without missing a heartbeat, went [*singing*], "Lift every voice and sing..."
>
> So, here we are, the only black people in the woods on the Fourth of July, singing at the top of our lungs "Lift Every Voice and Sing." And I was afraid, because I was thinking, "No, no! Shhh! They're going to get us!" On the other hand, it was so wonderful. "Yes, yes, yes! We are black people! This is our country, and this land is our land!" We should be able to stand in the woods and sing the black national anthem, you know? Own it! I don't know. I just feel like we've been robbed.

The relationship of blackness to land we think of as wild is always informed (but not wholly defined) by wilderness as an unsafe place. It remains inhabited by the specter of flight from patrollers enforcing ownership of black bodies and black labor. Every tree offers memories of fruits both sweet and strange. Still, through listening to black land narratives and through living my own, I know the wild is a place in which we belong and that belongs to us; it is also a physical space and an interior condition of which we have been perpetually robbed. At the base of a mountain or the edge of a city, the black wild is a state of mourning and awe.

## *Meaning in the Margins*

Land is not just a metaphor. It is where we plant our feet and where Earth's gravity tells us which way is up. Land has its own uses for itself. It continuously self-defines in river and bank, forest and glade, prairie and stand of oak. Blackness is the marker of those who were forced to redefine land's self-determined uses into cultivated and urban landscapes yet forbidden to make their own places there. Where, then, do black peoples' relationships to land live?

As people across the globe live more and more in cities, black people are already the most urbanized people in the United States. The enduring and resilient nature of black relationships to land force every person on the planet to remember that we can commit acts of dominance over nature, but that we are all beings unseverable from the wild. As my ears fill with stories of black people describing their existence in the natural world, they do so without distinction between "urbanized," "domesticated," or "wilderness." I hear deep knowledge of country living from people born in the heart of industrial cities. I hear narratives in which every space reserved as "wild" is already inhabited with a thicket of cultural meaning. They describe black relationships to land as an ecotone: a transitional margin where different ecosystems meet. In nature, ecotones are places of abundant diversity that generate life.

This understanding of relationship to land does not hold the center of our conversations about how to create resilient and regenerative relationships to nature, and yet it is thriving out here in the margins. This "margin" is not only the problematic social marginalization of those kept by violence away from the center of what is necessary to survive. It is also the ecotone at the margins of road—the ragged boundary between the solid, paved highway of empire and its edges, where black-eyed Susans grow.

This margin is not stark but soft, permeable, and fierce. It is not discoverable by the dead-eye whiteness of settler colonialism; it is alive in brown bodies who move through the geologic landscape of city boulevards, where waxy chicory and Queen Anne's lace defy playground asphalt to blossom. Black people teach our children to name these as wildflowers, even when other people call them weeds. Black relationships to land are not newly cultivated in charitable gifts of raised-bed gardens where people have temporary permission to grow food on land they do not own. They have always been present, thriving in poor soil, in patches of backyard collards waving themselves into sweetness in the weak October sun.

A black relationship to land can assume high-laced, Keen-brand hiking boots and topographic maps. It can also assume the casual way my sandaled feet know the slope and block-by-block topography of 125th Street, from Adam Clayton Powell Boulevard down to the Hudson River. This less well-shod relationship to land has been pitied in the rooms where American thought leaders are growing a new land ethic. What is lost in these pity-filled conversations is the value of every black land story, including my own.

Like most descendants of the Great Migration, my roots are in the rural south. I grew up in a Northern industrial city. I only know the names of the flowers and trees that can lift and climb their way out of suburban drainage ditches: willow, sumac, yarrow, pigweed. I am at home wherever nature butts up against this paved and built concrete world.

My experience of race, too, is an ecotone: a soft, permeable boundary zone, moving between things other people think of as distinct and separate, with or without value. My skin and this wild earth are marked with four hundred years of choices about cooperation and competition and misunderstanding of difference. These signifiers written on my body and its memory are often seen as a corruption or an ellipsis in relationship between people and land, rather than as the stuff of which such relationships are made. So who I am, and what I know, is that the wild is alive and changing, and it can never be settled, and it will seek freedom whenever we believe it to be owned.

I am the recipient of these seeds of wild black knowledge—there is no nature out there, only the soft boundaries between who we are and what we are willing to know as true. There is abundant life thriving here, on the edge of long woodland walks and taking sudden flight from urban housing lots.

This is a way of being on land that many Americans long for. Some think they must "invent" it or, having witnessed it, want to appropriate it for themselves. But like land, the black imagination has its own purpose. Pushed to the edge of survival, it holds the seeds of futurity. In unlikely places, it reveals ways of knowing wildness, land, and the possibility of freedom. In these wild black margins live the truths we, every one of us, are longing to call home.

## Note

1. Whately, an African American professor at Tuskegee University, promoted small-scale, direct-to-consumer, regenerative farming as an agriculturally and economically sustainable business practice for smallholder farmers.

# 15

## HEALING THE URBAN WILD

*Gavin Van Horn*

We are all indigenous to this planet, this mosaic of wild gardens we are being called by nature and history to reinhabit in good spirit. Part of that responsibility is to choose a place. To restore the land one must live and work in a place. To work in a place is to work with others. People who work together in a place become a community, and a community, in time, grows a culture. To work on behalf of the wild is to restore culture. —Gary Snyder, *A Place in Space* (1995, 250)

## *An Urban Wild*

There is the nature we discover and the nature we recover. There is wildness and there is wildness. And sometimes, our own wholeness depends on the nature we attempt to make whole.

I need to explain. There is wildness where I'm sitting at the moment, a glade in Eggers Woods on Chicago's Far Southeast Side, surrounded by a mix of American elm, shagbark hickory, bur oak, and sugar maple trees, whose leaves flicker in the sunlight. It's the kind of place that offers a glimpse of Chicago before there was a Chicago, one of the many parcels that compose the nearly seventy thousand acres of the Forest Preserves of Cook

County. The preserves, along with parks, gardens, backyards, golf courses, cemeteries, and railroad rights-of-way, form the green threads that stitch wildness into the city's fabric, that make the city a living place.

Eggers Woods is also a cultivated place. Chicago is rightly known as an epicenter for ecological restoration and for its volunteer ecological restoration movement in particular. Beginning in earnest in the 1970s, small bands of concerned citizens adopted certain neglected forest preserve parcels as experimental worksites for bringing back rare and threatened native plant species. There was a lot of work to do, and these restoration groups suffered some lumps, but their overall success—which has grown with time—has been called a "miracle under the oaks" (Stevens 1995) and has become a model for other regions of the country.

Most of the sustained volunteer-led work, however, has occurred in Chicago's suburbs and in more affluent neighborhoods in the city. A city known for its "sides"—Far North Side, Northwest Side, North Side, West Side, Central, South Side, Southwest Side, Southeast Side—Chicago has a history that includes structural racism, economic redlining, and white flight. Like any city, Chicago bears its history into a visible present, and many African American neighborhoods, primarily on the West and South Sides, show signs of such disadvantages. For people in neighborhoods where priorities are keeping schools open and businesses functional, the opportunity to connect to the natural world, much less restore ecosystems, is not the first order of concern.

I'm thinking about such things because across the picnic bench from me is Henri Jordan. Henri is an advanced crew supervisor for Greencorps Chicago, and therein lies the exception to the rule. I have spent the last several months getting to know the people and work of Greencorps Chicago, a program of the City of Chicago that specializes in contractual landscaping and ecological restoration work.

Henri has been working for Greencorps for more than four years, and it's easy to discern why he is a supervisor. In a word, he has presence—the kind of presence that is communicated by the way he weighs my questions, sifts them through, then measures out his words from experience. Among his crew, he's known as "the chainsaw guy," the one who knows trees by their Latin binomials and can discern the healthy ones from the unhealthy. He has four apps on his phone for tree identification, tries to learn a new tree every day, and reads continuously. As he puts it, "Everywhere I go, if it's green, I'm observing it."

Another thing about Henri: he is an ex-felon. He has lived his entire life in Chicago. By his own account, he "grew up in the ghetto and was exposed to the vicious cycle of the streets." Like many who have been through the Greencorps program, for Henri, the job wasn't just a job. Greencorps was a catalyst to a better life. "Everything in your life is a process of becoming," he tells me. "The question is whether you become, and as humans we have a large say in what we become."

In practical terms, Greencorps provides technical training, environmental education, and professional certifications to its crew members, positioning them for job placement in a range of environmental occupations. But there's more to Greencorps than prescription burn certification and properly administering herbicides. Greencorps is an organization whose work is transforming parts of Chicago by increasing the resiliency of urban lands. Sometimes this involves removing what's moved in—notoriously fast-spreading and invasive plants and shrubs such as buckthorn, phragmites, and purple loosestrife. Sometimes this means selective cutting of abundant trees so that light can once again reach patient seeds, which in the absence of regenerative fires have been stranded in time on the forest floor. Sometimes this means replanting species long gone, with the hopes that those animals that need them, from insect pollinators to endangered birds, will return as well.

Henri wipes a hand over his brow and removes his white hard hat, the kind you might see on a construction crew chief's head. As he does so, it occurs to me that a new kind of construction is occurring here in Eggers Woods and throughout the city's forest preserves—a construction of relationships between people and land, an opportunity for the emergence of wildness. There are nuts-and-bolts aspects to working with Greencorps—skill building and the disciplines of showing up on time and doing good work. But there's also something deeper that Henri articulates: "You're interacting with something that has life in it. Even if it's not another human, it has life in it, [which] touches something on the inside that doesn't often get touched."

My conversation with Henri underscores that the longer one does this kind of work, the greater the appreciation for how a living world works, for the wild systems that define a place. When I ask him if his ability to identify different plants makes him feel more at home in the preserves, he answers unequivocally: "Absolutely. I feel connected. Words give things life."

And what about the word *wild*? "I come from the West Side of Chicago—that's wild. Chaotic.... A lot of people don't get *out* of those neighborhoods," Henri responds. Gesturing around us toward the trees, he provides a contrast: "I don't call it wild [in the forest preserves]. I call it nature. It's what the blocks in the hood should be like—peaceful."

Is the ecological restoration work of Henri and other Greencorps crew members a restoration of wildness in the city? It's complicated.

## A Complicated Wild

In conservation circles, *wild* is a kind of shorthand for *healthy*. Vast mountain ranges. Foaming rivers sluicing through canyons. Page through an Ansel Adams wall calendar from the Sierra Club: you won't find Chicago or any other urban area; you won't even find people. The health of the land and water and clear blue skies, we are led to believe, doesn't have much to do with human actions. In fact, the fewer of us, the better for the wild, or so goes one version of the environmental narrative.

There are other associations with the word *wild* that don't figure much in conservation circles. When I asked people from Greencorps Chicago what the word *wild* calls to mind, as I did with Henri, they talked of lawless *humans*, ranging from the historical Wild West to current conditions in the West Side of Chicago. Guns and drugs, the demand for heightened awareness, checking over your shoulder, knowing what street corner you are on: These associations don't have much to do with healthy land, and many times they indicate the opposite. *Wild* means out-of-control or unruly territory, broken glass, abandoned storefronts, razor wire–topped fences, and learning to avoid such places for the sake of bodily safety.

To my eyes, there was wildness where Henri and I were sitting. There was also wildness—different in kind—a few blocks away.

The South and West Sides of Chicago can be high-risk places to grow up. The majority of Greencorps crew members call these parts of the city home. Of the persons who come through the Greencorps program, looking for career training and a fresh start, 90 percent have been incarcerated at some point. The forest preserves don't figure prominently in their mental maps of Chicago neighborhoods, even when these preserves are nearby. Zach Taylor, the former project manager for Greencorps, told me he's probably heard "over a hundred times" the exclamation, "Oh, I grew up my whole life right here. I had no idea this [natural area] was here."

Curtis Moore, a young man I spoke to who is in his first year of Greencorps training, provides a case in point. He grew up "ripping and running" with gangs in Englewood in South Chicago. "I just couldn't leave the streets alone, as far as hustling . . . not thinking right," he says, and the forest preserves were "never on my map." Later in our conversation, when I ask him about wildness, he responds, "Wow. I can associate wild with a lot of things, like what's going on in communities all across Chicago. Just *uncontrollable*." He pauses to gather his thoughts. Knowing the rough contours of the project I am working on, he continues: "I might have a different definition than you. What I see in the neighborhood, as far as all the violence, [that's] the first thing that would come to my mind, it wouldn't be as far as something with the forest. That's why I can't use *wild* with that. I'm so at peace there, so it's not wild there."

Tyrone Ellis is another Greencorps crew member I met in his first year of the program. We spoke in the Greencorps crew's truck cab as his colleagues were felling ash trees and pulling up phragmites on a chilly day. He grew up on Chicago's Southeast Side and describes his former disposition as comparable to a confused Tasmanian devil, "just wild, just wild." As for Chicago, "just living in Chicago, period, is dangerous," Tyrone remarks. Pointing north, he continues the thought: "People who don't think so live *that way*."

While wildness carries these negative connotations for many Greencorps participants, the meaning of the forest preserves, augmented by hands-on experience, changes over time. The ecological puzzle pieces lock in place, and the crew members who become advanced crew members or supervisors soon read the landscape with new vision. Controlled burns, a common restoration practice, can be especially eye-opening. This is also a management tool that lends *wild* new connotations. I ask Tyrone if the word *wild* has changed at all for him, given his time in Greencorps. "Oh, yes, it has. Wildlife. Wildfires. And both of them is good. You need the wildlife, you need the wildfire." He then lists the many benefits of burning—nutrients released that enrich the soil, new plant growth, eliminating invasive weeds.

Another person with whom I spoke, Brenda Elmore, described her journey through Greencorps as one that was full of new experiences. Brenda is a Greencorps success story. She graduated from the program and got a job with a nonprofit conservation organization, Friends of the Forest Preserves, where she now works as an assistant supervisor. She's been with them for three years.

Such a career would have been inconceivable to her only a handful of years ago. "Growing up, I didn't know anything about nature," she tells me. On the South Side of Chicago, she recalls, "The only time we could go to the woods was if we were picnicking, and we couldn't go anywhere near the treeline.... To my knowledge, the only thing you did in the forest preserves was sell drugs or something you had no business doing." It didn't help that she was afraid of wildlife. "I would run like crazy from deer. If a snake was anywhere near me, I would scream so loud you could hear me for six blocks away." In short, she was led to understand, "Black people didn't do nature, especially on the South Side of Chicago." Her work at Greencorps changed her perception of the forest preserves from places of menace to places of beauty. She has since become an advocate for these areas of the city, with a particular focus on connecting people of color to the forest preserves. As she put it, black people in Chicago need to be involved because "this is our earth. We deserve to be here. We deserve to help protect and restore it."

This kind of restoration between people and place may be wildness by another name. One face of the wild is represented by extremities—chaos, disorder, independence at the expense of others, a kind of self-will that in the long run is self-defeating. This is represented in Greencorps members' associations of wildness with the Chicago streets. When the Greencorps members I spoke to talked about the forest preserves, however, they most often used words like *peace, serenity, relief, fascination,* and *clarity of mind.* This indicates a different—an inclusive, collective—form of wildness. As the poet and essayist Gary Snyder put it, "When an ecosystem is fully functioning, all the members are present at the assembly. To speak of wilderness is to speak of wholeness. Human beings came out of that wholeness, and to consider the possibility of reactivating membership in the Assembly of All Beings is in no way regressive" (1995, 173). Ecological restoration in the city is one attempt—perhaps always inconclusive and provisional—to ensure that all members of the assembly are present. Human beings "reactivate" their membership in the process of doing so. Might this be wildness—to feel like participants in something that exceeds our control, that transcends our daily stressors and assures us of our place in the Assembly of All Beings?

## *An Emergent Wild*

By training and empowering skilled leaders who are from communities that consistently lack the opportunities available in more affluent communities, Greencorps Chicago serves as a foundational step toward a broader sense of connection to the natural world. A common theme that emerged from my conversations with people in Greencorps was the idea of mutual healing. The land benefits, gaining a measurable amount of health, but the crew members experience positive changes as well—some subtle, some dramatic.

Tyrone, for example, the "Tasmanian devil," was a reluctant Greencorps recruit. He'd been in and out of the penitentiary four times, and his "mind was still in the streets." He chuckled as he recalled the day of his interview with Greencorps and how he deliberated over what he would say to a friend who recommended the program to him: "I was thinking of a lie to tell him so I didn't have to go." The idea of "picking weeds" dumbfounded him. "Oh, nah, with your hands?" he remembers asking.

But being out in the field changed this perspective. Tree identification has become a passion, refined with time and experience. His perception of the forest preserves has been transformed from an "abandoned place where people dump" to "a place that we need." This need has become personal: "I love it. When you get out here you feel a peace of mind. I go through a lot of things at home, but when I come out here it all goes away." He'd recently read research about hospital patients with views of the landscape who experienced quicker recovery rates following surgery. This mutual healing made complete sense to him. Compared to when he was incarcerated, he said, "My mind is like on a whole other level."

I asked Zach about whether he thought the healing impacts of doing ecological restoration went both ways—for the land and also for the people doing the work. He almost cut my question off with his response of "absolutely." To him, the discovery that you are part of cycles that both transcend and include you is especially important to people who have been emotionally or psychologically wounded. A person may experience hardships, but "the flowers are going to come out in the spring. You can point to all these different things that remind you why it's good to be alive. If you're helping create that space to remind others, that's a real positive thing." Sometimes it's a simple reminder. Zach thought about Owen, a Greencorps crew member who was "away" for sixteen years, and how it was a healing experience for him to crush mountain mint from a worksite and take it home to use as potpourri.

Not only are the people directly associated with Greencorps crews impacted. A ripple effect occurs as well. People I spoke with frequently noted the pride they'd gained in making their communities better places in which to live and recreate. Friends, family, and strangers noticed: a honk from a car horn with a friendly thumbs up, people recognizing the Greencorps logo on their trucks and work vests, the curiosity of young people who see photographs from the worksites on Facebook and Instagram.

There is also the impact on crew members' immediate families. "I lost my marriage, my family due to the streets," Curtis tells me. "I've done so much destruction to the neighborhoods, and now I'm able to come give back and make the neighborhood look good." Seeing friends of his locked up, not returning, made him worry about his own three kids having to grow up without a father. His kids now recognize the motivational changes in him and that his stress levels are down because he's "a part of something that's bigger than just [himself]."

Henri notes that all three of his kids recycle now, that he has friends who ask him to identify the trees they see in their neighborhoods, and that his daughter has her own garden on their home's back porch, which includes plants grown from seeds he brings home from work. Brenda continues to draw from her own experience to "change the myth of the forest preserves" as dangerous places. This is now part of her full-time job, but it began with her children, who were immediate witnesses to her personal growth. All her children have now volunteered for various conservation projects in the city; one of her daughters became a crew leader for the Student Conservation Alliance (SCA); and her son was recently hired by the SCA, even traveling to Washington, DC, to meet with US senators about conservation. "I took the fear that ran through our family and changed that fear," she observes.

Renewing these mutual connections might be all the more important in a city, where urbanites' dependencies on the natural world may be less apparent. "Connecting to nature will give you a better respect for life in general," Henri notes. "Without these trees out here, we wouldn't exist," he tells his crew members, "so take that in while you are walking and doing your work." Curtis echoes this, saying, "If there wasn't no plants, there would be no us. That's it right there." In regards to other animals, Curtis continues, "In the urban area, just to see all the life that's lived off the maintenance we're doing, that's good to me. . . . The animals that I didn't really pay attention to or didn't think anything about, I have a respect for everything that's out there now. . . . Before when I was younger it was like 'whatever,' but now I just focus on peace for everything."

Cultivating the wild can shift one's perspective about place—by understanding the city as embedded in a larger bioregion—but it can also alter one's perspective of time. As Henri puts it, "This was here before we were here. The land was here before the people were here, so why not get to know the land you live on, that you inhabit?" When he takes his youngest daughter to explore the local woods, "it's like an addiction." As they look for different kinds of trees and animals and follow the ways that water flows through these areas, Henri notes, "I want to know more. I want to see more."

The ripple effect I heard about from the people with whom I spoke—healing the land, being healed by one's interaction with the land, advocating to others in one's immediate family and beyond to one's community—has brought me to a more nuanced understanding of wildness. After the infliction of so many wounds, the healing takes time, for the land and for people. When I ask Tyrone if Chicago is a different place to him now that he knows the forest preserves more intimately, he tells me, "It's a gettin' better Chicago."

As Henri and I discuss the kind of Chicago he wants his daughter to grow up in, he talks about the need to involve more people in caring for their local environments. When that happens, urban nature isn't an abstraction. "Now it's part of you." One of the reasons that ecological restoration is important within urban areas is that it provides a hands-on and close-to-home appreciation for the wild beings with whom we coexist. This can lead to an understanding of urban areas as lifeworlds full of other-than-human ways of being, as places that we are responsible for shaping with the needs of other species in mind. Perhaps wildness, in this sense, is not something discovered so much as something that emerges—from relationships that become "part of you."

\* \* \*

On a June day when it finally felt right to say "summer" in Chicago, Zach and I drove to the Far Southeast Side of Chicago, a place where the city brushes up against the border of Indiana. Chicago's muscular past is evident everywhere in this landscape: in the steel ribcages that puncture the ground, the chemical brews awaiting remediation, and the channelized and polluted waterways—all altered to suit the needs of industry. After passing semitrucks and weaving past the Ford assembly plant, we detour down a two-lane street, pulling to the shoulder of Hegewisch Marsh.

Zach has come full circle in a way. He grew up in the Southwest suburbs, but because of his affinity for wildlands, he moved away from Chicago: "I didn't feel there was a nature experience I could have." Working in places like Hegewisch Marsh—a biodiverse 130 acres of wet savannah, prairie, forested wetland, and hemi-marsh that was once a degraded industrial dumpsite—helped him rediscover Chicago.

Zach got to know Hegewisch by leading Greencorps restoration crews there for four years. The area was a mess when restoration began. For decades, the marsh served as an official dumping site for toxic by-products and slag from steel production and as an unofficial throwaway area for cast-off car parts and unsavory materials. The hydrology was compromised, invasive plant species were rampant, and deep grooves from four-wheel-drive joyriders crisscrossed the property. The amount of work to be done, Zach tells me, elicited more than a few groans from crew members when they first laid eyes on the site. Curtis offered me a Hegewisch summary: "Woo, rough."

Hegewisch is still a place of frayed edges. Funds have shifted elsewhere, and Greencorps is no longer on the job there. Signs of neglect are apparent. As I walk beside Zach, he notes the invasives—phragmites, reed canary grass, thistle—that are reclaiming the trails his crew built and the marsh edges that they spent many days seeding with native plants.

He shrugs his shoulders when I probe about whether he's disappointed to see Greencorps' work undone, redirecting my question to the experiences of the crew members who did the work. That couldn't be uprooted. They'll take those experiences with them. New skills, yes. New knowledge, yes. But also new relationships to Chicago and the nature within Chicago. An understanding that they are part of the larger lifeworld of the city. He tells me about the bald eagles he once saw here, the great horned owls that made use of a red-tailed hawk nest, and he points out one of his favorite plants, *Angelica sylvestris*, which looks like a creation straight from the mind of Dr. Seuss.

I tell Zach that one conception of wildness is simply an acknowledgment that the land has a will of its own. Wildness can be in the city, a self-expression of the landscape amid, alongside, and with human enterprise. Wildness, in this sense, indicates the *unique expression* of a landscape like Hegewisch. Rather than a declarative statement, wildness is a question that begins a dialogue with the land: What does this landscape *want* to be, *if given the opportunity*? Zach perks up at this, noting that there is a correlation with this idea and the experiences of the people in the Greencorps program. What do these men and women want to be, if given the opportunity?

I've walked the trails in Hegewisch a few times. A decade ago, the soles of my shoes would have melted had I strayed into the wrong chemical soup. Today, tree swallows flash their white underbellies toward us while skimming for mosquitos that hover above the shallow water. Daisy fleabane thrusts yellow sunbursts toward the sky. A dragonfly—a big-eyed, caramel-colored meadowhawk—comes to rest on my finger. Not perfect, not pure, not pristine. But relatively wild. A cultivated wild that needs continued attention, demands human involvement, and can change us as we change the city.

## Acknowledgments

Special thanks to Zach Taylor, Henri Jordan, Curtis Moore, Tyrone Ellis, Brenda Elmore, and Andy Johnson for spending time in conversation with me and for all the good work they do.

## References

City of Chicago. "Greencorps Chicago." http://www.cityofchicago.org/city/en/depts/cdot/provdrs/conservation_outreachgreenprograms/svcs/greencorps_chicago.html.

Jordan, William R., III, and George M. Lubick. *Making Nature Whole: A History of Ecological Restoration*. Washington, DC: Island Press, 2011.

Snyder, Gary. *A Place in Space: Ethics, Aesthetics, and Watersheds*. Berkeley, CA: Counterpoint, 1995.

Stevens, Wallace. *Miracle under the Oaks: The Revival of Nature in America*. New York: Pocket Books, 1995.

Woodworth, Paddy. *Our Once and Future Planet: Restoring the World in the Climate Change Century*. Chicago: University of Chicago Press, 2013.

# 16

## BUILDING THE CIVILIZED WILD

*Seth Magle*

The truck smells rancid.

That's usually one of the first things I notice as I climb in and fire up the engine before heading off to collect data on urban wildlife. I use scent lures to attract carnivores to our camera traps so we can get their pictures and find out where they are. The lures are an aromatic mishmash of rotting meat stench and musky gland smells, with a sort of soupçon of rotten egg thrown in. I try to keep them in sealed plastic bags, but somehow the perimeter is always breached, and I spend all day trying not to notice I'm driving in a sealed metal box that smells like a slaughterhouse.

I learned to use these scent lures from colleagues who deploy them in forests and prairies. But I work in the city, and so I drive the truck slowly through downtown Chicago, looking for the first little pocket of habitat on my list. In urban areas, anything green is habitat. Actually, so are some of the brown and gray spots. This is a lesson we've been slow to learn, we wildlife biologists. The city's long been a place shunned and reviled by ecologists.

For decades, the popular environmental narrative in North America has been the same—human development marches along, obliterating all traces of wildness and biodiversity in its path, leaving a lifeless gray wasteland behind. Meanwhile, a small group of passionate, committed conservationists works to protect the last remnants of nature in little pockets of habitat all around the world.

It's a good story. And for a time, it may have approximated the truth. After all, isn't that why we built cities? Oases of civilization, keeping the dangerous and unpredictable creatures at bay. We are in here; they are out there.

But now, things have changed. We live on an urban planet. More than half of us live in metropolitan areas, and much more than that in the United States. By 2050, it's predicted that perhaps 75 percent of the world's population will be urbanized. The war, if it was a war, is over.

The truck rolls through the aftermath of this great victory. It's dark out, the sun is not yet up, but I don't much need the headlights, not with all these streetlights. Wildlife biologists always get up early. In most places, it's because they want to get their equipment set up before dawn, to be ready before the day-active animals begin to move around. Depending on their location and research sites, biologists may need to avoid the tropical heat or find nocturnal carnivores. My concerns are different. I want to avoid drug dealers. I don't think they'd be thrilled to know I'm setting up cameras in their neighborhoods, and I've come to learn they are not early risers. The city's got its own challenges.

So I'm watchful as we drive around. A mourning dove is getting an early start, I see. A rabbit bounds across the sidewalk in response to our approach. In the past, when people characterized cities as desolate blights on the landscape, they couldn't have meant that literally. Some species, after all, have always lived here. Rats, pigeons, and squirrels have been urban neighbors as far back as we have records. In fact, they've lived nearby so long that many people don't think of them as wildlife. But other animals have also adapted to urban habitats. Birds, some common and some rare, use cities for habitat and migration. And mammals, too—including coyotes, foxes, bats, and others—are making proverbial lemonade out of lemons, finding new ways to make a living on our urban planet. For a wildlife biologist, there's plenty to study. One just has to know where to look.

I remember an encounter from when I was a young child, walking through my favorite park at night. I spotted a fox, bounding through the grass. When I moved, no doubt clumsily, to follow her, she dove into a sewer grate. I went on walking, and then I turned around and saw her pop back out of a different grate behind me. I remember thinking, *I thought I knew this park, but that fox knows it better than I ever will.*

\* \* \*

It starts to get light out after we've set up cameras in three or four different parks. Naturally, now is when I have to pee. This isn't a problem my colleagues have. In the Australian Outback, or in the Rocky Mountains, you can take care of such things more or less anywhere. I'm standing next to Interstate 55. So work stops momentarily while we find a gas station. Another day in the glamorous life of an urban biologist, and a small reminder that we are going to have to adapt to work in places like this.

With the light comes people, too—joggers, cyclists, speed walkers, and others with morning routines begin to populate the parks and the forest preserves. They ask me what I'm doing, because I look a little weird, wearing ragged, dirty clothing and carrying a clipboard and some strange equipment. I'm obviously working, but on what? When I tell them, each and every one has his or her own urban wildlife story to tell. Usually they're inspiring, but some of the stories are sad. Sometimes our urban wildlife neighbors can have us tearing out our hair in frustration. Anyone who's fought with rats in her home or rabbits in his garden can attest to that. The conflict goes beyond domestic intruders—people hit wildlife with their cars, and such accidents kill thousands of people each year (and untold numbers of animals). From animals, people contract diseases, frightening ones such as rabies, Lyme disease, and West Nile virus. Sometimes, though rarely, these animals attack our pets, or even people. I can empathize with these tales. I recall a particularly vicious faceoff I once had with a raccoon who grew unusually attached to a discarded pizza box. The encounter wasn't pretty, or inspirational. It was scary.

Worse yet, I don't have easy answers to give these nice people. Not even any clear ones. The solution for endangered northern spotted owls is very straightforward, though very difficult: conserve huge tracts of old-growth forest. But how do you engage with a problem like coyote attacks on people's pets? This is the most common form of conflict between humans and coyotes (92 percent of all such conflicts in a Colorado study). There's a human dimension—get people to keep their animals inside, which means education, and outreach, and cultural understanding. There's a biological part—figure out what the coyotes are eating, and find out what makes dogs and cats so appealing. There's an ecological part—which neighborhoods are attracting pet-eating coyotes and why? And it's hard to sum this all up in a sixty-second discussion with an irate homeowner who's mourning a dead cat and who thinks you're just playing in the parks and wasting everyone's time.

But I try to give these concerned citizens some tips. In many cases, perhaps most, wildlife management in urban areas is really people management, and education will be needed every bit as much as hard science. Some things are pretty basic. I know that animals like foxes and coyotes that are fed by people are more likely to attack people or their pets later on, presumably in search of food. And yet people keep feeding these and other animals like Canada geese and pigeons. In areas with a high prevalence of disease like West Nile virus or raccoon roundworm, people should be aware and should avoid approaching or handling injured or dead wildlife. Actually, in general, people shouldn't try to touch wildlife, as difficult as it may be to resist when you're inspired by an animal sighting. I try to encourage forms of sharing that don't involve disturbing animals—take a picture, share it on Facebook or Twitter, tell your family and your friends. You can use those encounters to help connect people to their ecosystem and learn about nature, which is probably more useful, and more gratifying, than an attempted cuddle.

But these tips are a lot less than satisfying. So here I am, often in the cold, in a foul-smelling truck, working long days to set up wildlife cameras all over Chicago. It's hard not to ask myself, *Why the heck am I doing this?* Scientists have studied animals in cities in the United States since at least the 1970s, and in Europe for much longer. Aren't these studies good enough? Maybe I can settle back into my office with some hot chocolate.

Unfortunately, most of these previous studies are based on questions such as "What's up with foxes in London?" or "Why are there so many rats in New York City?" Typically, the research is concerned with one species, in one city, for a couple of years. That kind of work is helpful in finding very specific answers to small questions, but it can't really give us an understanding of urban wildlife as a whole. Urban areas are ecosystems, where every facet is connected. Coyotes arrive partly because rabbits are present. Rabbits follow the grass and the gardens we plant. Then again, so do the bees and the insects . . . and so on.

Scientists haven't traditionally thought of cities in this way—as complex, interconnected ecological systems, as "wild." But they might be just as or more complicated than tropical rainforests, estuaries, or any other natural biome. If the complexity is there—if different species are present, trying to find mates, trying to find food, trying to find shelter, if you could round a street corner and see an unexpected creature at any turn—what's missing? Why don't we already think of these places as wild? Urban animals, plants, soil, and water are interlinked, after all, and hard to tame. And they change, and adapt, right alongside all the combined facets of our human landscapes: culture, economy, sociology, urban design, architecture, and the like.

And so we need a new kind of research design, one intended for understanding more than one animal or plant, or even one city. We need to create a holistic view of how urban ecosystems function, how they interconnect, and how the behavior of wildlife relates to the complexity of cities. We need to look at our cities with a rigorous ecological eye, the same way early naturalists did in unfamiliar wild spaces. We tried to remove the "wild" components, but in the end we failed. We are left with something... else. Not quite like wilderness, but not quite cultivated either. One of the last great unexplored ecosystems is waiting right outside our doors. That's why I'm driving around the city, setting up equipment in little vacant lots and cemeteries, places that hardly anyone previously thought were worthy of study.

\* \* \*

On a golf course, walking along a trail, I hear a noise in a bush and I go to investigate. I'm not too surprised to see a raccoon bound out and shuffle away. Raccoons are some of our most common urban neighbors. They're smart, adaptable, and tough. They also cause a lot of problems.

So I start thinking, how do we *really* solve these problems? Take a simple example: raccoons taking up residence in people's attics. Our current solution is to trap the raccoons from the attic and move them elsewhere (or euthanize them as humanely as possible). Some basic studies of raccoon biology and behavior can help us build better traps, handle raccoons more carefully, and perhaps (if we want to) move them somewhere they're more likely to thrive. But all of this is a short-term solution. New raccoons will find their way into those attics, and the problem will begin anew.

Ecological thinking can get us out of this cycle. If we understand what attracts raccoons to certain neighborhoods—what types of houses, what types of plants, which other species—and then what behavioral traits lead them to select attics as habitat, we can build different houses and neighborhoods, and we can better arrange green space. We can make sure raccoons never (or at least less often) choose to go into the attics in the first place, because it doesn't appear advantageous to them to do so. We can build spaces that are attractive to the species we'd like to see in our cities and less attractive to the ones we don't. And slowly, we can make cities into places where we can conserve rare species, too. This is an idea called *reconciliation ecology*, and it opens up possibilities for cities that are a far cry from lifeless wastelands.

If I give the impression that I'm the only one doing this kind of work, rest assured, I'm not a visionary. Several cities are trying to institute long-term ecological thinking. Both Baltimore, Maryland, and Phoenix, Arizona, are designated Long Term Ecological Research (LTER) sites by the National Science Foundation and are advancing our knowledge about the interconnectedness of biological, physical, and social aspects of cities. The reason I drive around Chicago is because I'm working on setting up the world's most extensive long-term monitoring network for urban wildlife. This network incorporates more than 120 field stations, all across the greater metropolitan area, including downtown green space, forest preserves, suburban habitat, and even the rural outskirts. While the large spatial scope of studies like this is useful, the duration is even more important. We've been collecting data for five years now, including information on various mammalian carnivores, deer, bats, insects, and birds, and we intend to continue to do so indefinitely. We collect data in all seasons (most wildlife studies are conducted only in the summer), which means we have a sense of how wildlife modify their behavior seasonally and how these changes might influence conflict with humans. It's a lot of driving, and a lot of looking for gas stations. But it's worth it.

Already, these long-term studies are yielding useful results. I've learned, for example, that water is the most important factor in determining where opossums will be located in Chicago, whereas for coyotes, it's all about human foot traffic—they avoid it studiously. And now, when a park district employee, forest preserve biologist, or city planner calls me wanting to know how to bring more coyotes into a park, or how to deter raccoons, I can look through the data and let them know. In this way, I can help prevent conflicts before they even start. Or I can help conserve and protect rare animals in the heart of urban Chicago.

And what will happen, I sometimes wonder, if I succeed? If rare birds and mammals were found in the heart of urban centers . . . would that make these urban areas wilder? Would it be wilderness? Or, by trying to carefully determine which species persist, do I actually create a more tame, managed landscape?

*  *  *

I don't do this fieldwork in my stinky truck by myself. There's always someone with me, partly because I need the help and partly because of those drug dealers. Usually it's an intern. Today it's Jazmin. She's a very bright young wildlife student from Northeastern Illinois University. She's an experienced intern. I know this because she passed her first test. A black-crowned night heron pooped on her.

The black-crowned night heron is a state endangered species in Illinois. It's a colony nester and mostly eats fish. The adults are black and white, are about the size of a football, and have striking red eyes that make them look a bit prehistoric. Their largest colony in the state is located in Lincoln Park, just a few miles from downtown Chicago, in one of the most densely populated areas of the third-largest city in the United States.

We don't really understand why the herons (more than 250 of them as of 2014) have chosen such an urban space for their colony, but we can make some educated guesses. There's a pond near the colony, managed by the Lincoln Park Zoo. For years it was a typical concrete-lined park pond, used largely for paddleboats. But in 2008, the zoo decided to go in a different direction. South Pond was transformed—the concrete lining was removed and the pond deepened. Native prairie plants were brought in to replace the turfgrass. Fish and turtles were introduced, and an urban wetland ecosystem, now called the Nature Boardwalk, was created in the heart of Chicago.

It's only fourteen acres in size, and I have no doubt that if you asked ornithologists before it was created, they would have told you that night herons would never use such a small site, especially one surrounded by noise, light, traffic, and pedestrians and their dogs. But there the herons are, more every year, stubbornly making their own choices as wild animals are wont to do. With the night herons, we got lucky. We built it, and they came. It's not always this easy. More often we need to learn what exactly it is that these rare species need. Why don't they live in the city already, and can we change things so they will? In a way, we position ourselves as the ultimate keystone species in this new type of habitat, not just managing, but actively improving diversity and resilience. It won't always work—we'll probably never conserve Florida panthers or polar bears in urban centers. But sometimes we can, as the herons are happy to show us.

That's not to say there aren't challenges for them, and for us. We work with the Chicago Park District to try to keep people from walking below the nests, because the juveniles frequently fall down and dogs tend to attack them. Herons poop absolutely everywhere, which is a dilemma for both landscapers and the researchers who try to study them. There are other problems with the herons' choice of location. A few animals chose to build their nests above the red wolf exhibits in the zoo, which spawned an interesting discussion on the ethical implications of a federally endangered species eating the fallen young of a state endangered species. There's a lot of work to do, but so far, trends are positive. If this one colony is the beginning of a movement, if black-crowned night herons are becoming an urban-tolerant species, then they likely won't be endangered for long.

People like Jazmin are the ones who are going to make the real gains in urban wildlife research. And maybe folks like the ones with whom she and I spend our day chatting. The most important resource for conserving wildlife in cities, a resource that is unavailable in less-populous areas, is, of course, people. My team uses citizen science projects as educational opportunities to connect people to their urban environment and, simultaneously, to assist us in data collection for ecological research. In Chicago, we launched a website called Chicago Wildlife Watch (www.ChicagoWildlifeWatch.org), where anyone with an Internet connection can review our images of urban wildlife and identify the animals in the photos. We've also reached out, with help from young people like our interns, to local high schools to train students and teachers to monitor wildlife using all the same techniques and equipment we use. Projects like these can help researchers gather data to which they wouldn't normally have access while training the next generation of wildlife biologists and connecting people to these new forms of nature—and such projects are far easier to pull off in cities.

Those future scientists are going to have to learn so much. It takes an incredible number of different skills to study animals in cities. The basic wildlife biology toolbox barely scratches the surface. You need to understand economics, sociology, culture, urban planning. You have to be just as comfortable with people as animals. You have to understand a system that changes faster, and more thoroughly, than any other kind of natural landscape. I think Jazmin's up to it. Sometimes I wonder if I am.

As we drive farther out of the urban core, it's on the bike paths and the greenways alongside the train tracks and the canals where we really start to see a lot of wildlife species. The reality is that there's a great deal we can do, on both large and small scales, to help make our cities places where humans and wildlife can coexist. One approach is to increase "connectivity" and thus the ability for animals to move through urban areas in search of other habitat. This may involve creating networks of open space or sometimes long, linear greenways for wildlife along landscape features such as streams or rail lines. These kinds of corridors make great bike paths, too. The ability to facilitate the movement of animals is becoming all the more important now, as climate change forces entire populations to shift their range in order to stay within the average temperature range to which they're adapted.

When roads are under construction, it may be possible to build culverts, underpasses, or even overpasses to help wildlife cross safely. This approach has been used to great effect in Canada at Banff National Park and in Colorado to protect lynx. Better yet, if we can learn how animals move around urban landscapes and decide where to cross roads, we can target road crossing structures, signage, and other features to ensure crossings happen in the safest places, for both animals and people.

\* \* \*

I'm out in a forest preserve now as the afternoon wears on. I open a camera case and an avalanche of earwigs—hundreds, perhaps thousands, of them—pours out onto my hands. Jazmin giggles at my undignified squeak of surprise, and then we collect a few for identification purposes. It's not all about the big animals, the mammals and the birds. If cities are ecosystems, we need to think about all the component parts, and typically we know even less about the small things—the insects and other invertebrates, the frogs and snakes and lizards, than we do about the more noticeable wildlife. For the really small building blocks, the soil microbes and such, we usually don't have the first idea of what's out there. But our gardens attract pollinators, and clearly there are bugs that are specialized to reside in urban systems. Cockroaches, sure. But also fireflies, bees, and other beneficial insects that are an underappreciated part of our cityscapes. So while, embarrassingly, we used to say *eww* and shake the bugs out of our cameras, eventually we realized that they provide necessary data, too. Understanding comes, even if slowly.

Back in the truck, as the sun is sloping lower in the sky, I see signs of the first steps toward wild cities. I spot green roofs and green infrastructure, urban nature parks, certified backyard habitats. More people are starting to treat golf courses and cemeteries as wildlife habitat, too. These efforts can bear real fruit and help make our cities a new, important form of nature. Cities will never be wilderness areas, in the stricter definitions of that landscape type. They do, however, have great potential to be areas of relative wildness—places where our everyday lives are lived alongside those of other animals. Places where we can more directly apprehend the needs of other animals and our impacts upon them. Places where we can seek to reconcile our needs with theirs.

But that new nature needs new scientists, too. These unique ecosystems will be different in every city. They will need careful monitoring, and engineering, in order to meet both human and ecological needs. This sounds difficult because it is a staggeringly complex problem—and an opportunity as well. Yet we have to do this, because if we don't, we'll find more and more animals making their way into our cities despite us, doing damage, creating hazards, and driving a wedge between people and nature. But we should want to do this, too, because by creating smarter cities we build a bridge—we begin to tear down that metaphoric wall between cities and the wild, to acknowledge that they are, and should be, linked. And maybe, years from now, we will look around these city skylines, once the great villains of our ecology soap opera, and see a place that humans and rare wildlife are both eager to call home. A landscape in which humans more humbly value and respect, negotiate and care for, the agency—the wildness—of other animals. A civilized wild. A new ecosystem for a new world.

For now, the truck rolls on. We have a long way to go.

## *References*

Beatley, Timothy. *Biophilic Cities: Integrating Nature into Urban Design and Planning.* Washington, DC: Island Press, 2011.

Lincoln Park Zoo. "Urban Wildlife Institute Biodiversity Monitoring." http://www.lpzoo.org/conservation-science/projects/urban-wildlife-biodiversity-monitoring.

Magle, Seth B., Victoria M. Hunt, Marian Vernon, and Kevin R. Crooks. "Urban Wildlife Research: Past, Present, and Future." *Biological Conservation* 155 (2012): 23–32.

Rosenzweig, Michael L. *Win-Win Ecology: How the Earth's Species Can Survive in the Midst of Human Enterprise.* New York: Oxford University Press, 2003.

# 17

## CULTIVATING THE WILD ON CHICAGO'S SOUTH SIDE

*Stories of People and Nature at Eden Place Nature Center*

*Michael Bryson and Michael Howard*

The red-tailed hawk cruises high over the city, the broad blue expanse of Lake Michigan behind him. Below, a tangled landscape of apartment buildings and houses, roads and railways, vacant lots, and fragmented bits of urban forest run west to the far horizon. His sharp eyes rove past the fourteen-lane-wide Dan Ryan Expressway that cuts a long scar down the South Side of Chicago, separating neighborhoods from each other and moving tens of thousands of vehicles per day. Just a few blocks west of that obscenely huge road is a tiny green oasis, a rectangle of calm and quiet at Forty-Fourth and Stewart. He heads toward it with swift wing beats, then rides the thermals generated from the city's urban heat island effect, checking for the telltale movements of his next potential meal.

Michael Howard looks over his charges. More than three hundred young African American schoolchildren are gathered around him here at Eden Place Nature Center, a green sanctuary for both people and nature in Fuller Park, one of Chicago's most isolated, contaminated, and poorest communities. The kids gambol in the sunshine and outside air, reveling in the freedom of release from their classrooms for a day of hands-on learning about the world. They squish their feet along a wetland's edge, decorate pumpkins, take hayrides, pet farm animals, and yell out questions to their heart's content. Few, if any, have ever seen a goat in the flesh or had an ornery goose nip at their heels—and they are joyful and overwhelmed by the novelty.

Michael is an articulate and dynamic presence, a teacher who uses whatever materials and critters are at hand at any given moment to communicate his love for nature, to share an ecology lesson, to tell a story about the earth and our place within it. Suddenly, a few children notice some nervous movements among Eden Place's flock of chickens. As Michael recalls,

> Someone said, "Why are the chickens acting like that, Mr. Howard?" I replied, "Well, they usually only do that when there's something above them." And sure enough, there was a big red-tailed hawk above. He was just soaring. So I told all the kids, "OK, everybody, I want you to just look straight up. Don't look into the sun, just look straight up into the blue sky, and you're gonna see Mr. Hawk. He's circling." He was very low at the time, and so everybody's heads were up like this, and you heard this "Oooh, I see it, I see it, Oh!"
>
> I said, "Yes, that's a red-tailed hawk, and he's actually soaring on the thermals, the warm breezes that are blowing right now. You'll notice his wings are made so that all he has to do is keep them spread; when the wind blows over them, it's actually holding him up. He doesn't even have to flap. You see, he's not even flapping his wings, but he's able to bank and turn. See how he's going around in circles?" And everybody's goin', "Ooh, wow, Mr. Howard, wow! Is that a real hawk?" And I said, "Oh, yeah." But as they watched the hawk, he started spiraling up. And the kids said, "Mr. Howard, he gettin' smaller!" I said, "That's because he's going higher and higher." Everybody watched—and by the time they couldn't see him anymore, they started clapping. And that was great.

That brief yet dramatic hawk–child encounter tells us much about the manifestation of wildness in urban environments and the fundamental need for and value of contacting nature directly within our large cities, where steel and concrete dominate the landscape and children's access to healthy green spaces can be limited. Such encounters are often, as elsewhere in nature, random and unexpected—and they provide a source of joy and surprise, intrigue and edification, for humans young and old. On the other hand, that red-tailed hawk—a commonly seen species throughout Illinois but still a delightful presence in an inner-city environment where green space is at a premium—instinctively sought out Eden Place as a rich source of food within a desert of pavement. Hence the simple yet profound ecological value of even the humblest of urban green spaces. Habitat—home for both wild creatures and city-bound people.

This, then, is one of the many secrets of Eden Place, a small-scale urban space reclaimed from decades of abandonment and contamination. A place that welcomes both the tame and the wild, the human and nonhuman; that provides a theater for their meaningful interaction; and that stokes the innate wildness within each of us. Eden Place and the human stories it engenders are thus about many things simultaneously: rebuilding positive connections to nature, facilitating encounters with the wild (however it is manifested), building community one family and one neighborhood at a time, and healing wounds of racial and environmental injustice inflicted over centuries.

* * *

The Fuller Park community in Chicago is an unlikely setting for a natural oasis like Eden Place Nature Center. The smallest neighborhood in this city of 2.8 million people, Fuller Park has long been one of its most physically beleaguered. Bounded by freight and passenger railroads to the east and west, bisected by the perpetually roaring Dan Ryan Expressway, and beset by persistent poverty and sporadic violence, this 97 percent minority community was for many years best known for two things: illegal dumping of waste, hazardous and otherwise, in its many vacant lots and pervasive lead contamination within the soil and residents' homes.

Just over twenty years ago, Michael and Amelia Howard moved back to Fuller Park and Amelia's childhood home in search of their roots and driven by their dedication to community advocacy. They didn't start out with a desire to be environmental activists or restoration ecologists, but in clearing a nearby vacant lot of a two-stories-high mountain of illegally dumped waste in the late 1990s with the help of countless volunteers, they created Eden Place—a 3.4-acre urban farm and environmental education site with gardens, livestock, barns, compost piles, ponds, trails, and what Michael calls "reasonable facsimiles" of native Illinois ecosystems.

The early stages of an oak savannah and prairie restoration take up the north half of this refuge, the only bona fide nature center on the entire south side of the city. A small pond and cattail-lined wetland are found between the prairie and the farmyard. Modestly sized and brightly painted barns stand against the tall concrete embankment of the railroad that looms along Eden Place's western border. Exhaust-streaked trains, passenger and freight, clatter by at short intervals. Too often, freight lines stop and idle here, engines rumbling, diesel fumes thick in the air. Raised-bed gardens sport squash, beans, peppers, tomatoes, and herbs.

The Howards' ongoing dedication to this project in Fuller Park is a monumentally unlikely and remarkably inspiring tale of environmental reclamation and neighborhood redemption—one that in its early days sparked violent threats from gang members who were upset at the encroachment upon part of their turf but quickly garnered the support of neighborhood citizens young and old. Now a valued and respected institution with two locations in the Fuller Park neighborhood—the original nature center and a more recently acquired three-acre farm extension site several blocks south—Eden Place's mission encompasses growing food, providing natural habitat, serving as a safe haven within the community, fostering environmental awareness, and fighting urban blight and decay by dint of its very existence. As Michael puts it,

> Eden Place puts out tentacles because it's a doorway to nature. That's our mantra: the doorway to nature on the South Side of Chicago. And so we put out tentacles and allow urban people to travel on those tentacles to experience the different levels of wild that exist—from the Boundary Waters to the State Fair to the Shawnee National Forest, from the Cook County Forest Preserves to the Wisconsin Dells. We put those tentacles out and make these roads that families and children travel in what we call fun yet educational experiences—we're about education, but we do it in a way that's fun because of how that ingrains on the heart. We're trying to build stewards. We try to show the importance of sustaining these different wilds that Eden Place exposes these families and youth to so that we can start building stewards in this urban environment that is so disconnected.

By cultivating an ethic of stewardship through a basic appreciation of the natural world, Eden Place opens doorways to youth and their families to explore nature outside of, even far from, Chicago. At the same time, though, these experiences of camping, hiking, canoeing, and exploring provide new perspectives that impact their lives in the city. Eden Place itself, as a constantly evolving green space, provides opportunities for putting a newly acquired environmental ethic of care into practice, whether that means tending a community garden plot, feeding the livestock as a volunteer, or teaching kids about the value of wetlands.

\* \* \*

Lamont was an eight- or nine-year-old boy who became a regular at Eden Place one summer—not as a formal member of a camp youth group but as a neighborhood kid coming over on his own, looking for something to do, tacitly seeking encouragement and support. Living with his cousin's unstable family, he experienced a turbulent home environment that summer, and thus Eden Place became a daily refuge for him, as has been the case with many children over the years. One day, as Michael related, Lamont came up to him brimming with excitement.

> "Mr. Howard, Mr. Howard—there's one of them dinosaur birds down at the little pond!" I said, "Dinosaur bird? OK, you gotta show me this." Because I hadn't seen a pterodactyl in a long time. So we went down to the pond, but I didn't see anything. At the time, I just played it off and said, "I guess he just flew away." The next day, though, Lamont came back and said, "Mr. Howard, that bird back there!" I said, "OK, let's go see that bird"—and so I went to look, and there it was, standing there among the mulberry trees we used to have down by the little pond; but it flew away so fast I only got a glimpse of it.
>
> The next day I went back looking for this bird, because from the beginning I didn't really believe the young man's description. This time I got a better look at it, and it looked like a juvenile great blue heron—small, wasn't as large as he was gonna get, and he was in fact down at the little pond. So I had to go back to Lamont and apologize. I said, "You know what? He *does* look like a dinosaur bird! You got great observation powers." And so throughout that summer all he did was identify birds. . . . That bird stayed around for a little over a week until he just finally disappeared.
>
> What was interesting to me about this young kid is how that one moment of observation turned him on, you know—he really just wanted to be able to recognize every bird so that he could come back and tell me the name. And so for the rest of the summer, that's what he did: he started auditing the different birds, and I gave him some pictures and a journal to write down his sightings. His math ability improved because we showed him how to calculate the number of sightings and the percentages of each species and put this into a computer spreadsheet. We're always trying to take every day and translate it into a greater learning experience.

We don't discount the potential of our kids at Eden Place, because we see them do so much on their own that amazes us. I'm able to recognize this potential, and if you give them the tools and show them how to use them, these kids are phenomenal.

While Eden Place provides a space for many such encounters with the nonhuman world, it also inspires us to reflect on and challenge what we mean by "the wild" in a contemporary urban landscape. "When we created Eden Place," Michael said, "the thought was this: if we build it, the wild will come." And so it has over the last fifteen or so years. Red-tailed hawks. Migrating songbirds. Raccoons, opossums, skunks. White-tailed deer seen in the damp mist at two in the morning.

Here, despite the nearby expanses of concrete, ubiquitous air pollution, and roar of train traffic, pigeons and red foxes forage for food and shelter, and nature manifests its power in various guises—from the weeds tenaciously sprouting up in sidewalk cracks to the lone coyote hunting for stray chickens in the farmyard. Fine distinctions between wild and agricultural, or between native and nonnative species, matter much less in this context than the provision of many potential points of contact among humans, plants, and animals. To a child who has never gone camping, or been on a farm, or gone to the zoo, or hiked in a wilderness preserve, the sighting of a hawk or a head-butt from a frisky farm goat is a meaningful first encounter with the wild.

*　*　*

During a visit to Eden Place, one is immediately struck by the people—their quickness to welcome, their warmth and fundamental kindness—even more than the visually arresting landscape. Eden Place is first and foremost a *human* oasis, a space in which people of all ages, races, creeds, and abilities can converge in an economically struggling neighborhood in the heart of Chicago's South Side to find conversation, fun, joy, a little music, mud to step in, meaningful work for body and mind, a shade tree, a cool drink, animals to watch, and pets to feed. Eden Place cultivates friendship, camaraderie, and laughter amid the earnest, honest work of growing healthy food and stewarding an urban ecosystem.

The Howards build this sense of community by offering a staggering range of events and activities—seasonal festivals, nature walks, gardening workshops, camping trips, children's day camps, school field trips—but also by providing time and space for impromptu gatherings or solitary reflection. Just by dint of its presence, as a place for anyone to come and explore—five days a week, rain, shine, or snow—Eden Place brings folks together under an umbrella of acceptance, something Michael and Amelia realized in a new and powerful way one spring. As he reflected,

> We were having an Earth Day festival the third or fourth year of Eden Place. It was a beautiful spring day, things were going good, and I was moving around, the staff were doing their thing at the booths, and then I noticed this lady crying. She was standing there with her hands over her face, crying. I immediately went into panic mode and said, "What's the matter? What's wrong? What's happening?" She didn't answer and just shook her head. I said again, greatly concerned, "What's wrong, what can I do to help?" And she said, "Oh, no, no. No, no. This is good tears." I said, "OK, you scared me. What's going on?" And she said, "Look!"
>
> I turned around, and there was this man, dancing with his daughter to the music of the DJ, spinning her around. I said, "Yeah, that's nice." She replied, "No, you don't understand. He don't play with his kids. This is the first time he's ever played with his kids. It's not that he don't love 'em; it's just that he always hard-working and then he's too tired. I'm so glad we came here today because this is the first time he's really relaxed and everybody's just havin' a good time and you have such a great place." And she started telling us how great it was that this space has helped her family just be able to relax.
>
> That's when I started understanding that Eden Place was considered a place of safety. It was a safe haven. That's why a lot of parents will let their children come here, because they feel—and this is their words, not mine—that "Mr. Howard ain't gonna let nothin' happen at Eden Place."

* * *

Wild spaces are often viewed as sites of healing, of reconnection, of sanctuary from the tumult and stress of the modern world. In national and state parklands, wildlife refuges, and wilderness areas, through monumental efforts that most Americans now take completely for granted, nature has been preserved in various degrees of wildness. These ostensibly "natural" but nonetheless carefully managed lands serve as important repositories of biodiversity by providing wildlife habitat. They also function as places in which humans privileged enough to afford the time and expense of travel may encounter said wildlife, mostly in rural surroundings remote from urban centers where the vast majority of Americans now live. But these senses of the wild—as a state of being, a place to visit, or a mode of recreation—constitute a primarily white, middle- to upper-class understanding of how one experiences an increasingly commodified (and thus threatened) nature. This is the wild that is defined, in part, as devoid of human presence—even when this perceived absence is more fantasy than reality.

Such a sense of the wild, though, is not necessarily shared by African Americans—particularly those who experienced, as Michael notes, "the disconnect from the land that took place because of the Great Migration"— who grew up in cities and left behind their rural Southern roots. The Howards fully recognize and are sensitive to this perspective, which at times serves as a conversational stumbling block in their interactions with some African American children, teens, and adults who have no interest in, no connection with, but plenty of ingrained fear of wild nature or who understandably associate working the soil with, as Michael says, "the horror stories of slavery, about people escaping through swamps and forests," and the subsequent disenfranchisement of Jim Crow society.

> All these people were running from the farm. . . . So for a lot of African Americans back then, the wild, the forest, the woods—that's the boogeyland. We don't want to go there, we've had negative experiences there. Our forefathers died in that swamp; our forefathers were tracked down and hanged from the trees in those woods. There was a long period in which African Americans in America didn't get any joy in the wild in the same way that everyone else did that came to America.
>
> When African American farmers were freed, they wanted land, they wanted a mule, and all the rest. The Jim Crow era, though, did a lot of damage by dissuading African Americans from making a living through farming, whereas the Europeans who were still coming in on boats to Ellis Island were given access to land. America gave away millions and millions of acres out West, and then they turned right around and gave up the land to the same Europeans but not to the ex-slaves. They set up the land grant colleges to help the farmers learn how to work the land, to teach them. They set up county agents as a resource of how to grow food and be successful at it. Then they provided low-interest loans so they could mechanize their farms. They were given all kinds of support so European Americans could succeed farming this great land and opening up this great space.

Michael reflects upon this history as we sit in his office on Eden Place's new three-acre property several blocks south of the nature center—an old industrial building with loading bays and vacant lots on either side. Covered with gravel, broken concrete, and the remnant snow of a recent January storm, the damaged soil of this extension site figures large in Michael's plans for expanding Eden Place's farming operation. He speaks earnestly, surely mindful of the ironic contrast between, on one hand, his vision of sustainable urban farming on Chicago's South Side that is interwoven with the goal of community economic development and, on the other, the disenfranchisement of America's rural black farmers, who only now have received belated recognition of their past ill treatment. "Just two years ago, they finally received some recognition in a lawsuit that went to the Supreme Court about the disparity with which the USDA has long treated black farmers in America," he notes, though that wasn't the end of the story. He continues:

> Though the suit favored the farmers who were named directly, it really wasn't a true reason to celebrate because it failed to recognize thousands of other African American farmers who had lost their land, whose families had been put off of the land, who were never compensated. Their heritage and the wealth that they lost as a family just dissipated, or were transferred to whatever county or state that took the farm for taxes.
>
> These types of injustices have forced African Americans to turn their backs on any kind of land value. We had land ethics, because we worked the land all the way back to Africa. We worked lands a lot of people would call arid or infertile to be able to support families and tribes. [In America], we were able to support families even on the plantations when they'd allow you to grow on some bad patch of land that they considered inhabitable for any real type of production; even there, we were able to scratch out enough to live.
>
> I actually witnessed that with my great-grandmother. She taught me how to pick cotton, how to hoe a row of beans. A sharecropper in Haynes, Arkansas, decades ago, this woman fed me and her for the entire summer on a dusty piece of ground that she fertilized with the mud from the pig pen and the chickens.

As Michael tells this story, his eyes light up with memories of his ancestors and the resourcefulness of his great-grandmother, and he notices my gaze falling on the row of hand tools (including a hoe) propped up in one corner of his office. Much of the labor invested in Eden Place's gardens and natural areas is, viewed properly, solar powered—an energy transfer embodied in those old-fashioned tools that teach lessons as well as turn the soil. Michael sees the connection between doing the labor of urban farming today and reclaiming cultural wisdom from the past. He admits,

> The youth of today don't know their history, and so therefore they're destined to repeat some of it. If they don't learn their history, though, they're gonna fall for the same rope-a-dope that their ancestors fell for—not so much fell for, but had to suffer through because of people's ignorance or indifference. And so it's very important that I teach a lot of history when I teach about the land ethic and the wilds of America that we have disassociated ourselves from. Urban African Americans need to understand that their forefathers were experts in the fields of farming, wilderness survival, and living off the land. They must acknowledge that if it hadn't been for that expertise of their great-great-great grandfathers and their great-great-great grandmothers, they wouldn't be here today.

This expertise is embodied in the knowledge and land ethic of Michael Howard, his family, and their dedicated team of volunteers. It is expressed in the related acts of healing the soil of Eden Place and transforming the spirits of its human visitors. Eden Place serves as a site to repair—one person, one child, one family at a time—the damage of several centuries' worth of cultural dissociation from the natural world.

*   *   *

It is difficult to adequately assess the ecological impact of projects like Eden Place on Chicago. From a bird's-eye view, a 3.4-acre nature preserve/farm/education center is a mere scratch on a very large urban palimpsest, a place arguably too small to significantly enhance the city's overall biodiversity or even mitigate the urban heat island effect of the Fuller Park neighborhood. The true impact, perhaps, is measured person by person, in human terms. As Amelia Howard has often said,

> Not everyone who comes to Eden Place will become an environmentalist. But if they become a lawyer, they'll become a green, environmental-minded lawyer. If they become a doctor, they'll be more homeopathic because they'll learn about the full nutritional standards that food coming out of the ground supplies to your body naturally, and how if you build a good foundation in your children's bones and in their blood when they're young, it'll protect them when they're older. Whatever you do, what you learn here at Eden Place will help you be better at it.

Restoring nature to a glorious past is not the raison d'être of this site, which features, at its best, "reasonable facsimiles" of Illinois's native ecosystems. But by taking a less-than-a-mile hike from the ear-shattering noise of the Forty-Seventh Street stop on the Red Line elevated train to the peaceful and surprisingly quiet borders of Eden Place, one can immediately recognize that human restoration is its real and perhaps most valuable purpose. The Howards know this: no matter their racial identity or prior ideas about nature before arriving at Eden Place, people will come away thinking differently about themselves and their role in the world after visiting such a place.

And what of the wild, both around and within us? How do reclaimed inner-city green spaces like Eden Place cultivate wildness? Michael Howard's mantra is a deceptively simple truth: "If we build it, the wild will come." Not just refuge-seeking animals, the furtive coyote, and the soaring hawk, but also the teeming microbes and invertebrates of a restored topsoil enriched by years of compost and rooted with prairie forbs and grasses. These, too, are manifestations of the wild. Not just the acquisition of ecological literacy among children and their families, but also the renewed joy of working the soil, tending vegetables and greens, planting trees—small acts of stewardship and healing that, bit by bit, counteract decades of alienation, fear, and cultural trauma. These, too, are reclamations of the wild through both bodily labor and cultural memory. At Eden Place, the wild is not simply *there*; it is always in the process of being remade.

# 18

## TOWARD AN URBAN PRACTICE OF THE WILD

*John Tallmadge*

In 1990, Gary Snyder published a book of essays with a provocative title, *The Practice of the Wild*. I had recently lost my teaching job and moved to Cincinnati to take up a deanship. It had felt like going over to the dark side after a decade of studying nature writers and taking students into the wilderness. Moving about from New England and California to Utah and Minnesota, keeping one foot on campus and the other on the trail—it was a satisfying life. I had never dreamed of ending up in the Rust Belt, a thousand miles from the Boundary Waters and a thousand more from the High Sierras. How could one hope to practice the wild in a city?

Snyder's book developed his conception of wildness as a principle of self-organization, resilience, and complexity in living systems. But his title continued to tease like a Zen koan, fusing contraries. *Practice* suggested discipline, patience, order, spirituality; it promised a kind of serenity. But *the wild* still evoked violence, disorder, energy, and unpredictability; it promised remoteness and freedom, but also otherness, danger, and fear. How could you practice something like that without becoming fearsome and dangerous yourself? I resisted the phrase, squirming in its grip until, eventually, something began to relax and open.

Living in Cincinnati was a far cry from the life of adventure and teaching inspired by writers like John Muir and Edward Abbey. I too had wanted to climb mountains and get their good tidings or plunge into desert wastes in pursuit of a hard and brutal mysticism. And even if I couldn't live in the wilderness, I reasoned that I could still return on periodic migrations, like the Canada geese. But settling in the city thrust me deep into the welter of civilization, where the clamor of engines, advertising, hungry crowds, current events, and everything we call "the built environment" drowned out the voices of the land. My experience resonated with that of the philosopher David Abram, who had learned how to listen to animals in the jungles of Bali yet found himself going deaf once he returned to America. I remembered theologian Thomas Berry's warning that humans have grown "autistic" with respect to the more-than-human world. These signs were not encouraging. Yet strangely enough, something different began to happen.

Perhaps it began with spending more time outside with my young children. Kids are small; they live closer to the ground; they pay attention because everything is new. To work with them, you have to slow down and set aside preconceptions. Start looking at things from their vantage point and the world appears strange and wonderful; you begin to appreciate how Walt Whitman could have been inspired by ordinary grass, or Mary Oliver by a grasshopper on a summer day. It was not long before I noticed bird's nest fungi sprouting in the mulch under my neighbor's spruce, the lichens gracing his maple trunk, the fireflies that rose into his treetops on hot June nights. I counted twenty-four species of birds within two blocks of our house — not just familiar kinds like house sparrows and cardinals, but charismatic red-tailed hawks and great blue herons. The woods in the park buzzed with insects, not just crickets and dragonflies, but periodic cicadas that emerged in clouds after seventeen years. The soil in the garden bristled with worms, tiny crustaceans, protozoans, and bacteria, many, as I learned from E. O. Wilson, unknown to science. Things seemed to get wilder and more numerous as you went down the scale.

Driving through town revealed tracts of land in every stage of ecological succession — from bare earth, to weedy vacant lots, to maturing second growth and actual old-growth forest. The longer a place had remained undisturbed, the more diverse its life appeared and the wilder it felt. I started to think that perhaps wildness was as much an experience as a condition. Not only that, but on reflection, it seemed to be a characteristic of all living organisms and therefore of ecosystems as well. Indeed, as Thomas Berry concluded, it must be a fundamental property of the universe, which is the greatest self-organizing system of all.

As a young adventurer, I had always thought of wildness and wilderness as one and the same, and I understood both in terms of space rather than time. Wilderness was remote, sublime, and pristine, free of human beings and their destructive works. The writers I admired resembled Old Testament prophets who forsook civilization for the desert, where they communed with God and were fed with the bread of angels. This was an enviable situation: out there everything was holy and clear, but the catch was, they had to go back. Wilderness gave the prophets their mission, but they had to fulfill it back in the human world, where things were messy, ambiguous, and confused. God and angels and revelations were so much harder to come by in civilization, so much more hidden and blurred; no wonder the old prophets loved the wilderness. Likewise, in the environmental mythology created by writers like Thoreau, Muir, and Abbey, wilderness was paradise and civilization meant ruin. Wilderness could be destroyed but not created; once gone, it was gone for good. This was a tragic narrative in which humans were the bad guys, all but the noble and few defenders. I did not find much comfort there as I struggled with homemaking and parenthood. It did not seem a promising path to sustainability.

But things changed when I began thinking of wilderness in terms of time and tried looking at life from the perspective not just of kids but of other organisms, all of whom grew where they were planted and did the best they could at getting a living. Wildness was all about change, becoming, learning, and evolution; that's what it meant to be self-organizing, to be, in the current lingo, a "complex adaptive system." Wilderness was nothing more than a place with a superabundance of wildness, and therefore it could grow back; it could return, even if not in the exact same form. Back home in New England, the forests had all returned. My grandmother had built our summer cottage in a hayfield; now wild turkeys and deer strolled out of the woods and bears had been seen less than a mile away. And right here in Cincinnati, where a deer in the suburbs had made headlines three decades ago, they were now as common as squirrels. Thinking of wilderness in terms of time helped me envision a world where people and nature might live in mutually enhancing ways. Wildness and human work need not be opposed. People could assume a nurturing, restorative role in the manner of gardeners and stewards. This would no longer be a tragic but a comic narrative, as ecocritic Joe Meeker suggests, for tragedy ends in death, but comedy ends in marriage, which in this case would mean aligning our lives and works with the methods of nature.

Of course, this is easily said but perhaps not so easily done. We don't immediately notice the wild all around us because we are accustomed to thinking of remote and glamorous landscapes where the sense of wildness assaults you at every turn. Robert Marshall famously argued that wilderness stimulates every one of the senses and was worth preserving on that ground alone. Thoreau complained of losing touch with his senses, as if alienation from nature had put him into some kind of trance, and he prescribed walking as a cure. Returning to his senses meant not only a return to direct bodily experience but also a return to the reason and sanity that had deteriorated from too much indoor living. The journalist Richard Louv has recently documented a syndrome he calls "nature deficit disorder," where children spend more time watching TV or playing computer games than they do outside, with adverse impacts on imagination, self-reliance, and health. It is not hard to see how indoor life can abet the autism that Thomas Berry describes by creating controlled, uniform conditions that allow us to concentrate on culture, on the world of language, symbolism, pictures, music, and abstract thought that we ourselves have created. It breaks the spell of the sensuous and in so doing can cut us off from wildness, which is nothing more nor less than the felt presence and personhood of other beings manifesting their singular history and character as they intently pursue their own ultimate concerns.

Ed Abbey had declared that wilderness was not a luxury but a necessity of the human spirit. And indeed, something in us seems unvaryingly drawn to wildness. E. O. Wilson calls it biophilia. The philosopher Martin Buber argues that human beings find and realize themselves not in autonomy but in relation; he calls the relation of means to ends "I-It," an apt description of our attitude when we think of nature as a resource or challenge. But when we encounter beings in nature on their own terms and feel their vivid, wild personhood, we enter into what Buber calls a relation of "I-Thou." Such transformative moments, he says, become the foundations of our spiritual life; they are found by grace, not by seeking, and they remain ineffably precious to us. It may be easier to come by such moments in remote wilds after days on the trail have erased civilized routines and subdued the clamor of memory. But they can also be found nearby; you just have to work a bit harder for them. You have to make yourself ready. It's a matter of practice.

What we want, I realized, is not an autistic but a dialogical relationship with nature, which, in the city, is just *relatively* wild—that is, it's not as manifestly wild as Yellowstone but, overall, is still wilder than a parking lot. It's a matter of scale and degree. We may not have big, fierce animals, but we have lots of smaller ones, many of which are ferocious enough. In the city, too, the wild is manifestly *related* to us, not only by shared genes and evolutionary history, but also because urban plants and animals take advantage of environments that we ourselves create. Humans are, in effect, the keystone species. Others flourish by adapting to us, but that doesn't make them less wild; it just demonstrates their resourcefulness. We may enable, but we don't control. Just ask any park employee whose shoulders ache from chopping bush honeysuckle.

After the shock of moving to the city had worn off and I had begun to travel around town, noticing the circulation of wildness and its penetration into domestic environments, the prospect of a hopeful, comedic narrative began to take hold. I wanted a future worth looking forward to, where culture and nature would interact in mutually enhancing ways. The Cincinnati landscape had been bruised and battered by commerce and industry for two centuries, but I soon noticed some hopeful signs of engagement with the relative wild.

Parks were the most obvious venue, particularly Mount Airy Forest, a 1,400-acre tract reclaimed from eroded hillsides that once held pig farms. It was acquired by the city dirt cheap starting in 1911 and by 1930 had been largely replanted with native trees; the Civilian Conservation Corps moved in during the Great Depression and built trails, shelters, and lodges using the local limestone. Now you can hike for miles with no sign of the city apart from a distant murmur of traffic; deer abound along with hawks, squirrels, and other birds and small mammals; some of the beeches and oaks are more than a hundred years old.

For something a bit more "pure," you can also experience virgin Ohio Valley hardwood forest at Caldwell Preserve. In 1914, after the Caldwells had made their fortune in whiskey and logging, they donated their last 122 acres to the city, and today you can walk under huge beech trees; through pawpaw groves; and past clusters of giant white trilliums, celandine poppies, round-lobed hepatica, and other perennials characteristic of old-growth woods. Some of these modest-looking plants may in fact be older than the trees above them. Both Mount Airy and Caldwell flourish because of human foresight, protection, and ongoing nurturance; they are like small blazes marking the path toward sustainability.

I found other signs of hope in certain people I met. Every community has outliers who pay close attention to the natural world. One day, the Sierra Club sponsored an outing along Mill Creek, which runs through the city's industrial gullet. For two centuries it has been channeled, dammed, diverted, and polluted in every imaginable way, becoming, to a casual eye, the very antithesis of wildness. Yet Stan Hedeen, a biology professor at

Xavier University, had spent thirty years studying its ecological history, and Robin Corathers, a local activist, had organized a movement to restore it for recreation. Stan led us from the headwaters through parks and industrial installations all the way to the sewage plant, where Mill Creek empties into the Ohio. There he had discovered a colony of black-crowned night herons, one of only two in Ohio; the chain link fence around the plant protected their nests from coyotes and raccoons. In Stan's eyes, the creek appeared as a living museum of interactions between nature and culture and a resilient, self-organizing system with an astonishing ability to cleanse itself once we stopped dumping pollutants; it was a worthy object of study and a tremendous resource for education. In the eyes of Robin and her activist colleagues, the creek was a neglected opportunity for creative restoration, healing, and recreation, a chance for Cincinnati to write a different kind of story on the land.

But the most inspiring example came from Fernald, a former defense plant just west of town that used to produce uranium fuel cores for nuclear reactors. During the Cold War, Fernald employed more than two thousand people under tight security, occupied ten huge buildings, covered more than a thousand acres, and shipped "feed materials" to Hanford, Washington; Savannah River, South Carolina; and Oak Ridge, Tennessee. Nuclear weapons were meant to keep the peace through "mutually assured destruction," or MAD in the acronymic lingo of the day. How we ever managed to live through that perilous time will be debated by historians, and perhaps theologians, for years to come. But in 1990, as the Cold War diminished and the Soviet Union neared collapse, Fernald was shuttered and decommissioned. It was well known that six years earlier an accident had released tons of uranium dust into the air, contaminating nearby land; the Centers for Disease Control and Prevention started monitoring townspeople and former employees. As cleanup began, local residents tested their water and discovered a radioactive plume in the aquifer. They pressed for a total cleanup, which ended up costing more than four billion dollars before it was finally completed in 2006.

Now the site has been made over into a nature preserve with restored wetlands and prairies that attract flocks of migrating birds and clusters of quiet hikers. At the LEED-certified visitor center designed by architecture students from the University of Cincinnati, you can learn about the complex and sophisticated engineering required to produce nuclear weapons. Grainy black-and-white photos evoke the desperation and resolve of the Cold War, when great nations prepared themselves to destroy the world in order to make their point. In these images, both technology and national power appear seductively sublime. The physicist Freeman Dyson once confessed, "I have felt it myself, the glitter of nuclear weapons" (qtd. in Schaffer 1985, 157). So have we all. And as recent events have shown, both nations and terrorists still lust after them. But a pillar of mushroom cloud can never guide humanity to the Promised Land. Fernald betokens another path, a new paradigm grounded not in the technological sublime but in the method of nature. It invites wildness to return and settle; it invites the people to ponder, and learn, and improve their practice.

Like any emergent process, this kind of promising work has a spontaneous, even haphazard, feel. What would it be like to practice the wild with more deliberation? Let me suggest beginning with five disciplines. Call the first *mindfulness*, a simple resolve to open our eyes, wake up to wildness, and shed our preconceived notions about urban nature. Mindfulness means devoting ourselves to the task of achieving a way of life that relates to the more-than-human world in honorable, dignified, wise, and healthful ways, for sustainability is not a state but a process. It's not something we achieve and then we're done; it's a perennial task, a lifelong work, a continuing relationship, and therefore a matter of practice. We can survive only by surviving and flourish only by flourishing.

Next comes *attentiveness*, by which I mean continuous learning from and about the wild. Simply put, it means studying natural history in both our schooling and our personal lives. We need to become ecologically literate. Without this discipline, we will never break out of our environmental autism; we will never learn the language of other beings. Natural history can renew our sense of wonder and mystery in the living world; we can draw inspiration from writers like Ann Zwinger, John Burroughs, Mary Oliver, and Henry David Thoreau.

But mere observation is not enough; we must also practice engagement with other beings. I call this *husbandry*. Food activist Michael Pollan suggests that a garden ethic offers more hope for sustainability than one based on wilderness, because it offers a productive space for human work. In my own efforts to raise vegetables, I have learned that it's more about self-improvement than about good taste or saving money. Aldo Leopold suggested that we plant gardens to avoid the spiritual danger of not owning a farm. Thoreau planted beans and harvested an instant and immeasurable crop; "It was no longer beans that I hoed, nor I who hoed beans," he wrote. When I planted squash, I soon realized that they were raising themselves, fulfilling their own inherent squashiness; I was merely facilitating that expression. In any event, both squash and I benefited: I ate their surplus flesh and they ensured the spread of their seed. Husbandry simply means dancing with wildness. And it's not just about gardening but about agriculture writ large and about ecological restoration, which requires us to learn how to dance with a multitude of species that offer no tangible or immediate payoffs in the manner of our customary crops.

A fourth discipline I call *pilgrimage*, meaning a devotional return to places of initiation and inspiration, the places that shaped our identity and our dreams. We need to honor our own Yosemites, Waldens, and Boundary Waters. We need to experience their refreshment and energy. We need to ensure that national parks and wilderness areas remain intact and inviolate so that our children and their children can learn what we know.

And finally, we need to practice a discipline of *witness*, by which I mean teaching and storytelling. Only witness, with its true stories, can preserve and transmit wisdom and knowledge in ways that we and our children can trust. As Barry Lopez says, "Sometimes a person needs a story more than food to stay alive" (1990, 60). Our stories are like medicine; they do their work only when passed on. Our stories are given to us, and if we don't give them away, they can't do their work; they fester within us; we can become sick at heart. Teaching, likewise, is a generative practice and a discipline of giving. It relies on stories. And it helps us live in hope, even if we feel exiled, even if we have to live in the city.

Now on my walks through the neighborhood, or even farther afield to Mount Airy Forest or Fernald, I rejoice in the return of wildness even in small, unobtrusive forms. Even a damaged landscape like Cincinnati holds abundant, enriching lessons in how nature and culture can dance together. John Muir said that he used to envy the father of our race who dwelt in Eden, but he did so no more, for he realized that the world was still being made, that this was still the morning of creation. We can, and should, still love grand, inspiring places like Yosemite. But we can also, if we choose, embrace the city as a venue for creative design and engagement with nature. That would be a wild, sustaining practice indeed.

## *References*

Abbey, Edward. *Desert Solitaire: A Season in the Wilderness.* New York: Simon and Schuster, 1968.

Abram, David. *The Spell of the Sensuous: Perception and Language in a More-than-Human World.* New York: Pantheon, 1996.

Berry, Thomas. *The Great Work: Our Way into the Future.* New York: Bell Tower, 1999.
Buber, Martin. *I and Thou.* Translated by Walter Kaufmann. New York: Scribner, 1970.
Hedeen, Stanley. *The Mill Creek: An Unnatural History of an Urban Stream.* Cincinnati, OH: Blue Heron Press, 1994.
Leopold, Aldo. *A Sand County Almanac.* New York: Oxford University Press, 1949.
Lopez, Barry. *Crow and Weasel.* San Francisco, CA: North Point Press, 1990.
Louv, Richard. *Last Child in the Woods: Saving Our Children from Nature Deficit Disorder.* Chapel Hill, NC: Algonquin Books, 2005.
Marshall, Robert. "The Problem of the Wilderness." *Scientific Monthly* 30, no. 2 (February 1930): 141–48.
McPhee, John. *The Curve of Binding Energy.* New York: Farrar, Straus, and Giroux, 1974.
Meeker, Joseph W. *The Comedy of Survival.* Tucson: University of Arizona Press, 1997.
Muir, John. *Our National Parks.* Boston, MA: Houghton Mifflin, 1901.
Oliver, Mary. "The Summer Day." In *House of Light*, 60. Boston: Beacon Press, 1990.
Pollan, Michael. *Second Nature: A Gardener's Education.* New York: Atlantic Monthly Press, 1991.
Schaffer, Ronald. *Wings of Judgment: American Bombing in World War II.* New York: Oxford University Press, 1985.
Snyder, Gary. *The Practice of the Wild.* San Francisco, CA: North Point Press, 1990.
Thoreau, Henry David. *Walden.* In *The Portable Thoreau*, edited by Carl Bode, 258–572. New York: Penguin, 1975.
———. "Walking." In *The Portable Thoreau*, edited by Carl Bode, 592–630. New York: Penguin, 1975.
Whitman, Walt. *Leaves of Grass.* Brooklyn, NY: self-published, 1855.
Wilson, Edward O. *Biophilia.* Cambridge, MA: Harvard University Press, 1984.

# PARt 4

*Planetary Wild*

# 19

## THE WHISKERED GOD OF FILTH

*Rob Dunn*

Brother of the blowfly . . . no one gets to heaven / Without going through you first.
—Yusef Komunyakaa, "Ode to the Maggot" (2000, 10)

Sixteen years ago, my wife and I, along with our friend Audrey, were standing outside a guesthouse near the towns of Boabeng and Fiema in Ghana when Kojo, a young boy, approached on a bicycle. His whole shadow rose and fell with each turn of the crooked front wheel. Behind him were miles of fields and a dust-dry forest. He stopped in front of me and opened his hand to reveal a small crumpled note. I unfolded it and read, "My friends, two of my children have died, i.e. the black and white colobus monkeys. Please come, quickly!" The note was signed, "the chief, Nana."

In the conjoined towns of Boabeng and Fiema, two species of monkey are considered by many to be living gods—fuzzy masters of the universe. As with any god, the relationships people form with the monkeys are individual. Some treat them with absolute reverence. Others scold them like misbehaving but well-loved children. But one man's god is another man's devil. Evangelical Christians in the town next door see the monkeys as false gods, impostors. The evangelicals sneak into the forest and kill the monkeys both to discourage the worship of false gods and to eat them. Apparently, sacred monkeys taste like chicken.

On this particular day, a taxi had hit and killed two monkeys as they tried to cross the red-dirt road into town. In addition to evangelicals, cars are one of several features of modern West African life with which these sacred monkeys can come into conflict, others being agriculture, logging, and hunger. In Boabeng and Fiema, when monkeys die, their burials include simpler versions of the ceremonies reserved for humans. It was one of these funerals to which we had just been invited.

As we approached the scene, perhaps a hundred people had gathered beside the road where one black-and-white colobus monkey lay in each of two small wooden coffins. Women were wailing. The chief was making an announcement. Other men were sounding their approval of the chief's words, a deep and mumbled assent. Overhead, the rest of the black-and-white colobus monkeys looked down.

Once upon a time, the forests inhabited by black-and-white colobus monkeys stretched in broad bands across West Africa. The colobus monkeys do not like to run on the ground, and in those days they could have flung themselves, tree to tree to tree, until their arms tired. These monkeys, like many of the canopy-dwelling monkeys of Africa, evolved a lifestyle that works well when canopies are expansive. But such freedom, if one might call it that, is a thing of the past. Today, the forests have contracted. If you could have seen us from above on the day of the funeral, you would have seen the bright colors of women's dresses, the red of the road, and the brown-green of the forest trees around us. But from a greater distance, something else would be conspicuous. The forest the monkeys live in is isolated, an island on which the gods are hemmed in, an island surrounded by fields of subsistence, fields where cassava and other stubborn tropical crops are grown to sustain the secular bodies of women and men.

Sacred groves conserve wild elements of the landscape by making those elements, those wild threads, sacred. The forest of Boabeng-Fiema is not the only sacred grove. Here and there across the country, here and there across West Africa, survive other patches of forests or tall grass in which living gods roam. The groves differ in their particulars, but all are living temples in which wild animal deities are worshiped through the daily prayers implicit in leaving a tree standing, resisting the temptation to shoot what is edible, and occasionally participating in more elaborate ceremonies, like the one that we had been called to attend. Beyond West Africa, sacred groves are also found. There are sacred groves in India, Japan,

Thailand, and Malaysia—thousands of them in total, each helping to save what remains. The precise relationship between the sacred beings and the humans differs among cultures in much the way relationships among humans differ, with details that matter, yet with core features as consistent as our human needs for food, water, relationships, and meaning.

Sacred groves are religious places to those who live alongside them, places where prayer of many sorts is offered. To the extent to which such prayer is a connection, it is a conversation between people and some not-quite-measurable value to the wild around them. That the wild is often relative (a patch of trees rather than a wilderness) does not change the significance of the connection. "Here," people say, "we find gods." Here, it might also be said, they find a living, growing reminder of something that is still beyond their control, fate, hopes, and dreams, which is embodied in the actions of the monkeys, bats, and even trees themselves, all the undomesticated forms of existence at the edge of town.

The groves, then, are spiritual connections to wildness. But in nearly every case, they are also functional, a traditional means of conserving the forest, swamp, grassland, or even just a small clump of trees on which both wild species and humans depend. Their sacredness preserves plant medicines, wood, and everything else the forests are needed for, a means encoded in religious taboos. As a model for conservation, the groves are not perfect. The forest around Boabeng and Fiema is a small, green donut where there were once miles of stubborn, sun-eating trees. But were there no monkey gods, things would almost certainly be worse. The last tree would have been chopped down, the last monkey eaten; the godliness of monkeys and trees holds the farms back, even if only just barely and in inverse relation to the degree of temptation. And, of course, sometimes temptation is too much. Even the wildest sacred groves typically end up devoid of those plant species useful in treating impotence. Surely, the logic goes, the monkey lords understand a little indiscretion.

Since I left Ghana, distance has highlighted the similarities rather than the differences between the sacred groves and Western conservation approaches. In the West, we don't hold up the animals in, say, Yellowstone, as gods. Yet the reverence we offer pandas, tigers, and other flagship species seems to me very similar to that given to sacred species. When we spend millions of dollars on a panda, it is probably fair to say we are valuing it as though it too were a god, a god whose conservation helps save the habitat in which it once thrived.

But recently, I've come to reflect again on the sacred groves. One of my former PhD students, Nyeema Harris, now a faculty member at the University of Michigan, has been doing conservation research (on big cats and giant rats) in Ghana. Another student in my lab, undergraduate Justin Hills, recently returned from a summer program in Ghana. In talking to both of them about my time there, I have had a chance to remember the ways in which sacred groves really are unique.

Although the word "grove" suggests a forest, sacred groves include other habitat types where species such as crocodiles, bats, and termites are protected. Not far from Boabeng and Fiema is a sacred river where the gods are fish. When we lived in Ghana, I was eager to see fish gods and so asked where we might find them. It is easy to imagine a sacred fish as being, in some way, even less predictable, and hence more magical, than a monkey. You can always find the monkeys if you look for them, but the fish, I thought, rise up in the water or disappear of their own volition. The symbolism of the sacred fish is rich and old. It is also biological. Fish, whose biology is mysterious and hidden by murky waters, provide sustenance and, when their populations wane, also take it away. To search for these particular fish, we were directed to a market in the city of Techiman. The river runs through the market, and so right there, in the middle of town, the sacred fish can, it was said, be observed. The prayer to the wild can be offered to blessed, shimmering fish.

After walking through the market full of loud colors and sounds and a rich diversity of pungent, sweet, and/or acrid smells, we came to the spot. There it was, a small pool of water. I was excited, a bit giddy even. Then I looked down. The water in which sacred fish bathe, I imagined, should be clear. This water was dark with waste. It stank of human filth and rotten market food. Yet there, rolling and twisting in the shallow vulgarity, were catfish, their long whiskers rising out of the water. Christians sometimes talk of a punishing god, but this looked like the reverse: a handful of gods being punished. One wants to find a lesson in such an experience. I think there is one, but it escaped me until recently.

One might see these fish as a tragedy, a measure of how capable we are of poorly treating even those we believe to be gods, how conclusively we are capable of extinguishing the light, even if relative, of wildness. That is how I saw the situation all those years ago, but in pausing to talk to my students about Ghana, I reconsidered. The fish are also something else. Here in the middle of the city were catfish living on society's discards. Catfish have always lived on what is left behind. They feel around in the dark for decay and, with their soft mouths and hard teeth, turn that decay into flesh. This, as I look back, was what was really different about the sacred groves: not the protections they afforded for the godlike monkeys, for which we have analogs in pandas and other charismatic megafauna, but instead the catfish. In them I met a belligerent god of rebirth, a god in whose body everything becomes wild again.

In Western cultures, we do not care for decomposers. We do not care for the species that turn what we do not want—the bodies of our relatives and pets, our waste, the waste of our society—back into life. The vulture is ugly, naked-headed, and repugnant. The maggot is worse than the filth it is capable of turning into wings. Then there are the termites, whose clear bodies raise dead wood onto legs and walk away. Most terrible of all, most utterly repulsive, the bacteria and fungi can bring nearly anything we give them back to life, but not before turning it foul, stinking, grotesque. Even the Bible, with its themes of rebirth, disregards the decomposers. Noah puts all the animals on the ark except, it seems, the decomposers, who were left to their own devices to spontaneously generate from filth.

The lesson from the sacred groves is that even what is ugly can be sacred and valuable, a connection to the wildness we most need. If the monkeys all disappeared, we would lose many things—perspective on ourselves, dispersal of valuable fruits, beauty, grace, the occasional funny story about a field biologist getting poop flung at her—but if the decomposers were to disappear, we would lose even more.

For the Egyptians, the dung beetle was holy. Historians note that the Egyptians thought the dung beetle was godlike because, when it rolled balls of dung (in which mother dung beetles put their eggs so that their larvae might feed on fungi and bacteria), it looked as though it might also be capable of rolling the sun. But let's not allow the story of the dung beetle's sacredness to obscure the beetle's practical value to the Egyptians. Just as the catfish at the edge of the market eat what we wish would go away, the dung beetles made more beetles out of excrement. Until we understand all of life, can manage every piece, which at our current rate of study will be never, we rely on these wild forms to do our work. In a time before modern sanitation systems, it is hard to think of anything more miraculous. These organisms return us, whatever our sophistication, to life; they rewild our waste and our bodies. Bless the beetles for what they do; bless the catfish and the vulture, too. Each one is, I can see clearly enough now, a kind of god. Take away the monkeys and we will suffer. Without the decomposers, nothing that we produce would ever be reborn.

Among the wild things, you can choose what seems holy, godlike, a god even. More and more, one sees patches of forest and ponds, remnants of some other era on landscapes composed primarily of the meaty slabs of farms. It is easy to imagine these remnants and their isolated birds, squirrels, or even monkeys as sacred, a wildness that connects us to something larger. Many books, many writers, have considered these patches and the children who wander through them unaware that they have discovered anything less wild than the landscape in which we evolved. We should cherish these patches, should worship and expand them. As for me, though, I'll take the bacteria and fungi. I'll take the vultures standing on rooftops and fences, raising their angular wings as if in some evolutionarily choreographed tribute to Martha Graham. I'll take the dung beetle. I'll even take the maggot.

Anybody can celebrate a monkey, panda, or squirrel; they are easy gods to love. They elicit a straightforward sort of worship, one of fences and nature reserves, the worship of the wild that remains. The decomposers are more difficult to deify. But their presence is consequential: They are everywhere and they need to be, for without them nothing would be reborn. Nothing would ever be wild again. Without them, we would all be knee-deep in feces and bodies. Without decomposers, even the plants would eventually stop growing. Because of them, instead of filth, we drink water. Instead of old grapes, we drink wine.

Some gods are clever, some gods are beautiful, some gods—it has been said but not proven—are even merciful. You can have those if you want. As for me, I'll take the maggot and the vulture. I'll take the bacteria. I'll even take the catfish rolling in the shallow stink of Techiman's market, the catfish whose groping mouth reaches up like the afterlife, forming a tunnel through which, as the poet Yusef Komunyakaa reminds us, we must pass to get to some other side.

## *Reference*

Komunyakaa, Yusef. "Ode to the Maggot." In *Talking Dirty to the Gods*, 10. New York: Farrar, Straus, and Giroux, 2000.

# 20

## THE *AKIING* ETHIC

*Seeking Ancestral Wildness beyond
Aldo Leopold's Wilderness*

*John Hausdoerffer*

"What kind of ancestor do you want to be?"

Michael Dahl's question hangs in the thick Minnesota air.[1] I lean against the deck railing as the chill of evening tightens my skin. Two stories below, Round Lake reddens, matching the maples that are flashing their last burst of fall color. Michael, White Earth Anishinaabe and Midewiwin Lodge leader, offers me the intense grin I suspect he gets when he's just floored a person with an idea. This makes me like Michael Dahl.

"What kind of ancestor do I want to be?" I repeat Michael's query slowly, quietly, as if talking to myself. The question implies that we are, always and already, ancestors. Even before our descendants are born. Even if we never have children. In terms of *space*, nowhere is, ethically speaking, "away." In an age of globalization, we grapple with how our actions impact peoples and environments on the other side of the earth. Now Michael asks me to think about ethics in terms of *time*. No era is "away." Ethically, all times and all generations are now. So how are we to live?

We are sitting on Winona LaDuke's deck. LaDuke—White Earth Anishinaabe author of influential essays and books like *All Our Relations, Recovering the Sacred*, and *Last Standing Woman*, as well as activist, founder of Honor the Earth and the White Earth Land Recovery Project, and former presidential running mate for Ralph Nader—clinks around inside her reclaimed wood house while Michael and I talk. LaDuke is making dinner for her daily entourage of kids, grandkids, neighbors' kids, interns (who live with her family), activists, and spiritual leaders. Her kitchen is the hearth of both a home and a movement.

One year earlier, I hosted LaDuke on my campus. On the drive home from the packed auditorium, she told me that my car smelled like horses. I took it as a compliment. That week, I had been teaching LaDuke's work in the same class that just read *A Sand County Almanac*, the mid-twentieth-century classic by American conservationist and author Aldo Leopold. My class noted parallels between LaDuke's recovery of "land to which we belong" and Leopold's call for changing "the role of *Homo sapiens* from conqueror of the land-community to plain member and citizen of it" (204). I, somewhat awkwardly, told her about my interest in pursuing these parallels as part of a new research project on whether the environmental justice movement could renew the relevancy of Leopold's famous 1949 "The Land Ethic." In "The Land Ethic," Leopold challenges readers to measure right and wrong, in all aspects of life, based on whether or not their actions as landowners, consumers, or citizens "preserve" the "integrity, stability, and beauty" of the "biotic community" (224–25). Summoning the courage to ask if I could interview her at White Earth about these connections, I wondered: *Would LaDuke think I was reducing the complexity of her ideas—ideas that emerge specifically from her people's time, place, and struggle—to those of a dead white guy? Am I any different than George Catlin, whose legacy my last book criticized for romanticizing the ecological views of Indians? How does an academic visit White Earth after so many academics stole the bones of LaDuke's and Michael's people and sought to patent their sacred rice?* These questions stilted my words as I struggled to explain my intentions for the visit. LaDuke, perhaps weary of my hand-wringing, cut me off with a two-word sigh—"Just come."

Now I am here, above Round Lake, having just driven twenty flat hours east from my university town in the mountains of Colorado. I am here because I suspect that LaDuke's efforts to "recover the sacred" can add new dimensions to Aldo Leopold's concern for the "spiritual dangers" of environmental alienation. Specifically, I suspect LaDuke's work can infuse multicultural and social justice realities into the ecological heart of Leopold's anxiety about how modern consumers are detached from the natural sources of their livelihood and the consequences of their comforts. I believe LaDuke's philosophy and practice of "recovery" can bring new solutions to address Leopold's fear that our daily ignorance of how we endanger living systems, in turn, endangers our potential as caring beings. After a week at White Earth, I found far more than a connection between Leopold's hope for a land ethic and LaDuke's call for native justice. I learned a view of land—*akiing*—that challenges, strengthens, and surpasses Leopold's ideas of land, wildness, and ethics.

Knowing Leopold's biography, I suspect he would be open to a concept like *akiing* challenging his view of land to evolve. Even when he wrote "The Land Ethic," as Curt Meine points out in the film *Green Fire*, he said that "nothing so important as an ethic is ever 'written'. . . . It evolve[s] in the minds of a thinking community" (225). *Evolve* is, in fact, a fitting word for characterizing Leopold's career. Historians celebrate him for changing his mind against major land-use practices that he once heavily promoted, from predator control to fire suppression, when faced with new evidence. His life (1887–1948) spanned the formative years of American conservation.[2] Trained in Gifford Pinchot's Yale Forestry School, Leopold worked in the American Southwest from 1909 to 1924 as a forester, game manager, watershed specialist, and recreational planner for the US Forest Service. During those years, Leopold's perspective evolved from utilitarian, timber-driven valuations of forests to asking land managers to consider "the whole loaf" (1923, 87) of "The Forest" (1913, 43). He evolved from viewing "The Forest" as a commodity reduced to board feet of timber to perceiving "The Forest" as a holistic, dynamic, and enlivening process underlying the natural world. Leopold expanded this view of the forest to include "trans-economic value"—the value of both ecological services and spiritual fulfillment that soil, water, grasses, wildlife, and recreation provide. By 1924, this vision inspired him to design the first wilderness area in the Gila National Forest of New Mexico. Over the span of fifteen years, Leopold's view of the land's wealth evolved to both imagine and manage land as something more than quantifiable resources—as wilderness.

## *Round Lake, White Earth*

Sitting on LaDuke's deck with Michael Dahl, the aroma of Round Lake mingles with the wetland scent of wild rice simmering on the stove, harvested from the lake. Solar panels surge and lights flick on as the dark rises. I tell Michael that ethical questions like his ancestor question are what I have dedicated my life and career to. I share with him what I know of Leopold's background and basic philosophy, quoting the passage that brought me to White Earth, to this deck above the floating rice: "There are two spiritual dangers in not owning a farm. One is the danger of supposing that breakfast comes from the grocery, and the other that heat comes from the furnace" (1949, 6). I elaborate: "Here's this ecologist, Aldo Leopold, who spent his life studying soil erosion and wildlife and forests . . . very material things . . . yet he chooses to open his great book with the phrase *spiritual* danger, rather than the danger of ecological collapse or the danger of lost wilderness."

Michael's eyes widen. "I like it. . . . That's my fear. If my people ever don't rice, if we stop manoominike ['going ricing'] . . ." He pauses, takes in the night, the season shifting to the Ricing Moon. His eyes focus on me again. "If we go to the grocery store and buy a pound of rice that we might have gotten off the lake anyhow . . . if we stop ricing to feed our family, then the ceremony of manoominike becomes abstract." For Michael, "spiritual dangers" occur when the spiritual becomes abstract, when manoominike becomes a ritual rather than a livelihood emerging from land shaped by ancestors who were reciprocally shaped by that very land, which in turn cyclically shapes us and our descendants.

The next morning, I will ask LaDuke about Leopold's concern regarding spiritual dangers. Without blinking, LaDuke will call it "the spiritual danger of not giving a shit . . . [and not giving a shit] hurts your spirit, because your spirit is connected to everything." As LaDuke's words rang in my head, I thought about how Leopold's ideas of the land's value evolved to encompass wilderness areas. Looking out at the sugarbush scrolling past the car window, thinking about how long that forest has been tapped, burned, and enlivened by Anishinaabeg knowledge, it struck me that as much as Leopold's thought evolved *into* promoting wilderness, Leopold's idea of what wilderness *is* evolved as well.

First, Leopold's purpose for wilderness evolved from recreational experience for hunters in the Gila in 1924 to a more scientific and ecological understanding of "land health" in places as far as Mexico's Sierra Madre in 1937. Yet ecological perception was not the ultimate end of his wilderness concept. Leopold began to shift his wilderness lens toward working and urban landscapes beyond "virgin" lands, all of which he felt revealed what Gary Snyder has since called the "wild process in all things" (2014). Even as early as 1925, in an essay titled "Wilderness as a Form of Land Use," Leopold asserts that "wilderness exists in all degrees, from the little accidental wild spot at the head of a ravine in a Corn Belt woodlot to vast expanses of virgin country. . . . Wilderness is a relative condition" (1925, 135). He continues: "As a form of land use [wilderness] cannot be a rigid entity of unchanging content, exclusive of all other forms, it must be a flexible thing" (1925, 136). Leopold would not go so far as to say that wilderness is a human construct in the way that many have suggested. Leopold thought that the "head of a ravine" (something we would expect to be "wild") has something in it that cannot come from us—a wildness that can only develop from "the remote fastnesses of space and time" (1949, 148). However, for Leopold, this wildness could develop in any sort of place: "The weeds in a city lot convey the same lesson as the redwoods; the farmer may see in his cow-pasture what may not be vouchsafed to the scientist adventuring in the South Seas" (1949, 174). By highlighting that studying weeds teaches as much and harms less than traveling to the great redwoods, Leopold imagined a relative wild.

## Egg Lake, White Earth

The day after my ancestor discussion with Michael Dahl, the bluebird sky above the White Earth Reservation has a clarity that only seems possible in early fall. I follow LaDuke's pickup truck. Her canoe is strapped to the top, while her thirty-year-old friend Gimiwan rides shotgun. We head to Egg Lake, just a few miles west of her house, for a day of harvesting wild rice. Egg Lake is the third lake we'll try today. The first lake had been harvested already, and the second did not have the healthiest beds. Familiar with LaDuke's essays about the sensitivity of wild rice, I worry the second lake has indeed been compromised by acid rain and climate disturbance.

We unload the canoes onto the bright clay soil from which White Earth gets its name. LaDuke and Gimiwan "feast" the lake with handfuls of tobacco. Gimiwan, drum in hand, cuts the breeze with high, soaring pitches from his usually low voice. I notice LaDuke looking down at the lake in prayer. I have grown accustomed to her dynamic energy, so her stillness focuses me. Later in the day, she will explain that she was contemplating her "covenant with the creator," her commitment to the instructions written onto the land from human ancestors and nonhuman relations that will enable her people to "hang around another thousand years."

LaDuke and Gimiwan slide the canoe into Egg Lake. I follow on a stand-up paddleboard, out of respect for the canoe as a sacred space for Anishinaabe. LaDuke sits in the front with two eighteen-inch-long batons known as "knockers," while Gimiwan stands in the back and poles the water. She sweeps the rice in with one knocker and taps the rice husk off the stalk with the other. The sounds of the canoe floating into the rice surprise me. It sounds like wind through dry-leafed trees, like rain on a metal roof. I want to hear all at once the chorus of sounds heard on these lakes over thousands of years, but I am tone-deaf when it comes to the echoes of ancestral instructions spoken through the land.

That night I will ask Gimiwan if those rice beds feel like "wilderness" to him, and he will reply, "Wilderness, to me, is home. My ancestors are in that soil, and every time I think about it . . . when we say our ancestors, we mean the animals, those are our ancestors, too. . . . Those are our family and we are thankful for having them because without them there would be no anything."

The canoe slices through the floating mud in which the rice grows. The mud creates its own universe, a microcosm of "The Land" that Leopold called "a fountain of energy" and that Gimiwan calls, simply, "home." Like the turtle's back described in Anishinaabeg origin stories, the mud drifts above the floor of the lake, forming a miniearth and microhabitat. Rice mixed into the sedge wetland undulates with the rolling water—a waving, moving earth. As the canoe works its way through, we see an eagle feather resting on the water. LaDuke gives me a handful of tobacco and asks me to pick up the feather and leave the tobacco in its place. Surprised and honored to be asked, I feast the lake, practicing give-and-take with the world that sustains us.

We paddle toward the shore of Egg Lake as the day of ricing ends. The bottom of the boat, filled with rice husks, looks like the backside of some massive porcupine. We fill white canvas bags, preparing our harvest for parching on the other side of the reservation.

Camping for the night beside Round Lake, my arms sore from paddling, I reflect that the idea of *wilderness* might have evolved at least once more had Leopold lived longer, more toward and perhaps beyond *wildness*. In "The Land Ethic," Leopold argues that "the capacity for self-renewal" (for a whole biotic community, including humans) defines "land health" (1949, 221). In turn, anything that leads to self-renewal defines health over violence and right over wrong. His celebrated "Land Ethic" is rooted in and measured by "self-renewal."

The root word for *wilderness*, scholars like Roderick Frazier Nash (2001, 1) and activists like Dave Foreman (1998, 404–5) have told us, derives from old English, meaning "self-willed land." Gary Snyder notes that the word *wild* means "self-managing . . . self-organizing, self-maintaining, and self-disciplining" (10–11). Thinking about Leopold's passion for "self-renewal" and LaDuke's struggle to recover the self-organizing Anishinaabeg traditions between ecological health and human livelihood, I wondered: How can "self-willed" communities like LaDuke's find common cause with movements for self-willed land that look to Leopold for wisdom? How can those communities, fighting against their place becoming everyone else's "away," join forces with wilderness efforts to restore the social and ecological "capacity for self-renewal" necessary to a thriving world? How can *wildness* be redefined to relentlessly demand "self-renewal," "self-will," and the capacity to "self-manage" for human and nonhuman communities worldwide? How can the concept of "self-willed" and "self-managing" communities unite rather than divide wilderness and social justice movements?

Driving out of the forest and onto the prairie of White Earth, LaDuke cautions me against drawing too many analogies between the Anishinaabeg lifeway and Western notions like wilderness or wildness:

> This whole question is a colonization of knowledge, this whole question. The taking it and the interpreting of it. Arguably, part of the [mental] process that everybody has is like "Okay, what this person is doing over there in this other world that I am really not familiar with . . . I am going to take it and interpret it so I understand it." That's what we have to do, because we are not that person, so we got to go through our own process. So it is one thing to do that, but that does not mean that it fits into a box.

Whereas Gimiwan views wilderness as "home," LaDuke does not speak of "wilderness." She is not necessarily against the term *wild*. She defends with her entire being the fact that "wild" rice should refer exclusively to the ancestral ceremony that feeds her people and her lake. Her retail business, Native Harvest, sells a poster celebrating ricing with the motto "Keep It Wild" in bold letters across the bottom. That said, to draw analogies between Anishinaabeg ancestral wildness and Leopoldian eco-centric wilderness (or even his evolved "wildness") carries colonialist dangers.

Struggling to shift out of habitual, problematic terms, I ask her about the meaning of the Anishinaabeg word *akiing*. LaDuke pauses, then answers:

> "Land to which the people belong." . . . It's also the land that we know . . . like, when I walk in these woods, I know what's going to help me. I know where my medicines are. So it's this relationship . . . it's a recognition of a mutual interdependence between you and the land. For example, I am doing

some scholarly work on the blueberries. It turns out that the blueberries respond best to fire. Three words for fire in Ojibwe: Ojibwe fire, white man fire, and fire caused by thunder-beings. And so the fire made by Ojibwe is for the purpose of bringing the blueberries, or bringing the medicines.... Another perfect example is that word *maashkode*, which is our word for "prairie," but in the etymology of the language, *maash* is the word for "medicine"... and *kode* is our word for "fire." So really the word means "the place of the burned medicines," which has to do with how you make prairies and has to do with the fact that a lot of prairie plants are fire plants.

*Akiing*, then, is both land to which the people belong and land that requires the people. This reminded me of my conversation with Michael, during which he told me that the practice of ricing also "grooms the lake." He said that the lake has more abundant wild rice because they harvest rice. The canoe and ricers provide a healthy disturbance—not unlike fire—that helps the rice to germinate the following year. Thinking about *akiing* as an alternative to both wilderness and wildness, I realize that *akiing* captures something that both of the former never will—*akiing* invites humans to become cocreators in addition to preservers of the "wild process." We cocreate blueberries and biodiversity when we burn, we cocreate healthy wild rice beds when we harvest, we cocreate the richness of the prairie when we rely on its keystone species—the buffalo. Thinking of the buffalo, I ask LaDuke, "So, just to be clear, the prairie therefore also needs a culture that needs buffalo, not just a buffalo management plan, but a culture that needs buffalo in order to be who they are?" She replies, "Right, and a people that sing for the buffalo, that drag buffalo skulls, that use every part of the buffalo." *Akiing* not only redefines the value of land or the human-land relationship; *akiing* recovers the Anishinaabeg place in the universe.

LaDuke's description of *akiing* cycles me back to Michael's question of ancestry. I struggle to imagine how I might be a good ancestor—more than a crumbling photo evoking positive feelings on my grandchildren's wall someday—without bequeathing to my community and cocreating with the land some understanding and practice of *akiing*. As hard and as dangerous as it might be to try to "cocreate" in a globally industrial society, I struggle to envision any other way out of Leopold's warning about the spiritual danger of not knowing the sources of one's food and energy.

Like Leopold, Michael Dahl's spiritual dangers come when we assume we can sever livelihood from spirituality and thus when we sever spirituality from labor and lakes and canoes and wild rice and blueberries and people and ancestors:

"I'm going ricing."... There's so many different pictures we get in our mind when we say it in English.... But instinctively, when I hear *manoominike*, I picture a lake, I picture a canoe, I picture a pole, I picture knockers, I picture that hide that lays on the bottom of the canoe when you've riced about 30–40 pounds of rice in the canoe—looks almost like a moose hide. I hear the sounds of rice, I hear the squeak on the side of the canoe, I hear the trickle of the rice hitting the canoe, I hear the popping of it when it's parching, I hear the popping of the fire, I smell the fire, I taste the rice... all encompassed in one word: manoominike.

I want my four-year-old son to grow up and not take for granted that not everybody gets to greet their dad when they come off the lake. Not everybody gets to lay on the garage floor and open a piece of green rice off of a tarp and eat it. Not everybody gets to parch rice and eat freshly popped rice. Not everybody gets to do that. I don't want my son to take that for granted.

I almost ask Michael what the "wild" in "wild rice" means in his language and where in the word manoominike might one find some idea of Anishinaabeg "wildness," some notion of what must be protected and enhanced in order to avoid Leopold's spiritual danger. But I don't need to. I understand that spiritual dangers, for Michael Dahl, are the dangers of one day not being able to ask, "What kind of ancestor do you want to be?" in the same way. If the rice beds, the ceremonies, the moose hide of rice on a garage floor for a child to play in, and the language that emerged from one's ancestors' livelihood are gone, then the ability to ask that question linking the empowering past, the urgent present, and the hopeful future disappears as well. The question itself becomes abstract—the ultimate spiritual danger.

I realize that before I can answer the question of what kind of ancestor I want to be, I need to understand *what my rice is* as a transient descendent of transient people long removed from the source of their food, heat, and ancestors (and as someone who does not love to grow food and whose once-broken back screams when splitting wood every fall). I realize that the future of wildness for me does not necessarily rest with the six wilderness areas that sit within an hour of my house. The future of wildness lies in the discovery of the self-willed human and nonhuman communities that make up my "rice" and in the *akiing*, the land to which my people (whoever they are) belong. I must, somehow, cultivate an "*Akiing* Ethic."

Someone calls from LaDuke's doorway—dinner is ready. I walk into the light of her home, brightened by the full darkness now outside. The heat inside lets the muscles of my chilled shoulders relax. I sit down for a meal of wild rice. It tastes like the lake.

## Notes

1. The scenes in this chapter reference events and statements made in the presence of the author and include personal interviews during and after ricing between September 12 and September 19, 2012.

2. This analysis of Leopold's wilderness evolution has been adapted from my 2013 essay titled "The Capacity for Self-Renewal," published in the September 2013 issue of *Minding Nature*.

## References

Foreman, Dave. "Wilderness Areas for Real." In *The Great New Wilderness Debate*, edited by J. Baird Callicott, 404–5. Athens: University of Georgia Press, 1998.

Leopold, Aldo. *A Sand County Almanac, and Sketches Here and There*. New York: Oxford University Press, 1949.

———. "Some Fundamentals of Conservation in the Southwest [1923]." In *The River of the Mother of God and Other Essays by Aldo Leopold*, edited by Susan Flader and J. Baird Callicott, 86–97. Madison: University of Wisconsin Press, 1991.

———. "To the Forest Officers of the Carson [1913]." In *The River of the Mother of God and Other Essays by Aldo Leopold*, edited by Susan Flader and J. Baird Callicott, 41–46. Madison: University of Wisconsin Press, 1991.

———. "Wilderness as a Form of Land Use [1925]." In *The River of the Mother of God and Other Essays by Aldo Leopold*, edited by Susan Flader and J. Baird Callicott, 134–42. Madison: University of Wisconsin Press, 1991.

Nash, Roderick Frazier. *Wilderness and the American Mind*. 4th ed. New Haven, CT: Yale University Press, 2001.

Snyder, Gary. "Keynote Address on the Working Wild." Presentation at the Headwaters Conference, Western State Colorado University, Gunnison, CO, September 19–21, 2014. http://www.western.edu/headwaters.

———. *The Practice of the Wild*. Berkeley, CA: Counterpoint Press, 1990.

# 21

## ON THE WILD EDGE IN ICELAND

*Brooke Hecht*

Picture a country hanging from the Arctic Circle, where at least 80 percent of the people leave room in their minds for the existence of elves, "Huldufolk" (hidden people), or other netherworldly creatures, where *wild* means vast stretches of grayness—gray, craggy mountain peaks; gray gravel; and gray ash from yesteryear's volcanoes.

If you imagined Iceland, you guessed right. And I was heading north into that gray.

I drove on roads carefully designed by the Icelandic Road and Coastal Commission, around bends that intentionally avoid the presumed dwellings and churches of elves and hidden people. I felt grateful for the possible company of these sprites, as I sometimes drove for hours without seeing another human, much less a gas station.

I had landed in Reykjavik the day before, ready for the first field season of my PhD research in ecosystem ecology. My bags were full of equipment that I hoped would help me unlock ecological puzzles about what made ecosystems flourish or fail. On the bus into town from the airport, a group of women—headed north for some adventure of their own—had clucked over me, concerned that I was venturing into this remote country by myself.

I wasn't as worried. There is really only one main road in Iceland, aptly called "Highway 1." Furthermore, I was looking for a forest—which I guessed would be hard to miss in this vast grayscape—a wild surprise of green life rising up from the ash.

I drove alone for two days, owing to the detour I chose to take around Snaefellsjokull. Known in these parts as the most beautiful glacier in the world, Snaefellsjokull is also the starting point for the expedition in Jules Verne's *A Journey to the Center of the Earth*. I was not seeking the center, however, but the edge: forest edges to be exact—the treeline, the forest limit. A place that might span a few steps, a place where you could throw a small stone through an ecological doorway, with you standing in forest and the stone landing in tundra. These mysterious threshold zones hold clues to what makes a forest a forest and what causes a forest to reach a breaking point and give way to tundra.

When I arrived at my destination, a forested valley in northeastern Iceland, I became less puzzled by the Icelandic belief in elves. The gnarled birches had an aura of magic about them. The drops of dew on the ancient equisetum (commonly called "horsetails" for their soft, feathery look) seemed to await collection into tiny fairy cups. What struck me most particularly was the scale of the forest. It was as if I had drunk from the Alice in Wonderland "grow" bottle as I drove. Either I was much taller than I had been two days before or else I had arrived in a miniature woodland, a fine home for elves, trolls, and perhaps a fairy ring.[1]

The sheep that dotted some of the woodlands seemed normally sized, from my vantage point anyway. Trios of sheep (a mother and two lambs) are a regular sight across Iceland in springtime. There are a number of interesting points about Iceland and sheep, and here are four: (1) there is a distinct breed of Icelandic sheep (note that there is one breed of sheep but thirteen different kinds of elves); (2) there are more sheep than people; (3) sheep run free, while forests are enclosed within fences; and (4) sheep have everything to do with Iceland's particular type of wildness.

A short time after arriving in Iceland, I learned that the sheep and, to a lesser extent, their grazing compatriots (goats, cows, horses, and pigs) have changed the Icelandic landscape so dramatically that the original Viking explorers might believe they'd discovered a new island were they to make landfall in Iceland today. When the Vikings arrived in Iceland, there were no native people—and no four-legged inhabitants except the arctic fox. The land was not gray but a lush green "from mountain to seashore," with "butter dripping from every blade of grass in the land" (Anon. in Jónsson 1946, 27).

In hindsight, it is easy to understand how newly introduced grazers would decimate the abundant vegetation that greeted the Viking explorers. As old trees fell, emerging seedlings were devoured, leaving little chance for new trees to grow beyond the reach of hungry sheep. Without vegetation, there was nothing to hold the soil underfoot. Where a horde of sheep had eaten its fill, the light volcanic soil simply blew away, leaving glacial till as the new terrain. The upshot was not so much a *novel ecosystem* as much as it was *ecosystem loss*, which comes with all the sadness that you might expect such a loss entails. This cycle (beginning with eating and ending with glacial till) continues in some parts of Iceland today.

If you know that Iceland had an influx of two-legged and four-legged creatures around 1,200 years ago, you also know that the present lack of color in many parts of this country is in fact a *loss* of color over this time period—and a story with many layers. I believed I would uncover important pieces of this story at the forest limit.

After spending one summer hiking every Icelandic woodland I could find, I honed in on three different birch sites. The forest limit zones of these three woodlands would become my study areas for the next few years. "Spend as much time there as you can," one of my dissertation committee members told me. "Drink your coffee there. Eat there. And just *look*." I don't believe I could have started my research with wiser words than those.

However, there are some things you cannot *see*. For example, as you hike to the forest limit—straight up—there are important changes in the leaves that cannot be seen with the naked eye. They need to be measured to be known, either with equipment in the field or back in the lab with carefully transported leaf samples. One such change that is important to this story is foliar nitrogen—the nitrogen in the leaves of the birches.

Imagine that you are hiking in a forested Icelandic valley and that you begin hiking up a mountain, through the forest and toward the forest limit. As you take each step upward, foliar nitrogen levels climb higher with you. There is an important reason for this. The higher up on a mountain you are, the colder it is. The colder it is, the harder plants need to work to photosynthesize (i.e., make their food). Nitrogen is key to a plant's ability to photosynthesize at low temperatures.

So high levels of nitrogen at the forest limit are essential to survival. This is a pattern that is observed worldwide—I can picture the pattern of the graph in my head—foliar nitrogen increases with altitude like steps going up a staircase.

There is something else important to this story that could not be seen as I took those woodlands in, day after day, over the course of several summers—and that is the woodlands' history. What had happened here over the last thousand or more years since the Vikings' arrival? What about the last few hundred years?

As an ecologist, I was painfully aware of the stresses that ecosystems worldwide experience from grazing, climate change, and other human-imposed stressors. What I wanted to know was this: Does a forest with a history of higher levels of disturbance have a more difficult time responding to additional stress than a forest with a lesser history of disturbance?

There was one way to find out. I would impose a disturbance on three woodland sites and observe the response. My three sites were strikingly similar birch woodlands, but they had a few important differences in their disturbance histories. My Site 1 (the forest in the valley in eastern Iceland that had me believing in elves) had not seen any serious sheep grazing for about a century. My Site 2, in a valley adjoining Site 1, was remarkably similar in all respects to Site 1, except that it had never been protected from grazing. My Site 3 was farther north—a harsher climate, a shorter growing season—and like Site 2, it had never been protected from sheep grazing. These sites were on a gradient of stress from the least stress (at Site 1) to the most stress (at Site 3). Knowing how important nitrogen is to plant survival at high altitudes (and latitudes), I would track foliar nitrogen as my clue, using it as my insight into how the woodlands were handling stress.

I didn't know at the time that some of the ecological models concerning disturbance, ecosystem shifts, resilience (or lack thereof), and crossing of ecological thresholds were based on psychological models of human psychic breaks and breakdowns. But now it makes sense. At what point does the accumulation of disturbances become so profound that a person—or a forest—is no longer able to function?

It is important to note that the prospect of disturbing the woodland sites was not an easy one for me. I was conflicted. I was studying forests because I loved them. Was it ethical to stress my subject and push it closer to the edge, even if my long-term goal was to understand (and even promote) ecosystem resilience? My advisor, Kristiina Vogt, comforted me: the forest disturbance would be minor and temporary. The ecosystems would bounce back.

With that reassurance, I bought a lot of sugar (actually, almost half a metric ton) for my disturbance experiment. While ecologist and forest service colleagues in Iceland questioned whether I was embarking on a homemade liquor and bootlegging project, the truth was that my unusually large sugar purchase had everything to do with nitrogen. A story from one of my fellow doctoral students, Michael Booth, can help me explain how.

Michael used to begin his forest ecology presentations with a p[...] a forest upside down. The roots of the trees were featured on top[...] leaves down below. His point? Much of what is running the show in [...] is under our feet. In any given handful of dirt, there are millions to b[...] of bacteria. And these microbes can be the tail that wags the forest dog, especially when it comes to nitrogen. While these bacteria play a key role in making nitrogen available to trees and plants in their preferred form, bacteria also need nitrogen for their own survival. Can you guess what happens to nitrogen in a handful of soil when there is a significant increase in the bacterial population? The answer: The microbes take the bulk of the nitrogen for themselves, leaving less nitrogen available for plants.

I wonder if a happy, healthy forest is one that has just the right number of microbes (whether that number would be in the millions or billions, I have no idea)—such that the microbial community gets the nitrogen it needs while giving the trees and other vegetation the nitrogen they need. While notions of "balance" in nature are very out of fashion, to say the least, the concept seems applicable here. Too few or too many microbes would be a problem—from the perspective of the Icelandic woodlands, anyway. At both ends of the spectrum, there would not be enough nitrogen for the plants and trees.

So what does sugar have to do with this? I could use it as a free source of energy for microbes—put enough sugar into a handful of soil and you might even cause a microbial population explosion. If I spread a bunch of sugar at my forest limit sites, where the birch trees are already at their threshold of existence, would the woodland sites with the higher levels of stress have a harder time dealing with it?

I spent quite a bit of time spreading carefully measured quantities of sugar in selected "disturbance" plots at my study sites while leaving an equal number of plots as controls (without the sugar disturbance). I carried sugar by the backpack-full up to the forest limit, ever thankful to have a wonderful field assistant to help me with the haul. We spread the sugar in the woodlands by hand. As my advisor had shown me, it's all in the flick of the wrist.

We spent even more time gathering birch leaf samples to bring back stateside to the lab for nitrogen analysis. I packed thousands of leaves for transport in a huge box. (In accordance with my permit to transport biological material across international borders, the large box was marked "quarantined material," which made a few of my fellow air travelers quite nervous.) I subsequently spent a lot of time in a basement lab at the Yale School of Forestry and Environmental Studies, dropping small capsules of

Icelandic leaf powder into a machine that resembled a clothes washer but was in fact a high-tech piece of equipment that would help me determine foliar nitrogen levels.

I was surprised to find that my study sites did not fall into the global pattern of increasing foliar nitrogen with increasing altitude. The location of the forest limit in terms of altitude was lowest at Site 1 (eastern "elven" forest protected from grazing), in the middle at Site 2 (eastern grazed woodland), and highest at Site 3 (northern grazed woodland). So I should have seen a stepwise increase in nitrogen levels from Site 1 to Site 3. In contrast to the expected trend of nitrogen levels climbing up as regularly as stairs, the nitrogen levels at my study sites dropped from Site 1 to Site 3 (with significantly lower nitrogen levels at Site 3 than either Site 1 or Site 2).

While this result offered me a new ecological puzzle right off the bat, my sugar disturbance shed some light on both this unexpected result and my original question about how sites with higher levels of stress handle additional disturbance.

At the eastern protected forest (Site 1), where there were lush layers of springy moss and no sheep, the sugar disturbance caused no change in the foliar nitrogen levels at the forest limit. In ecological speak, this site had "resistance" to the disturbance. However, at the two grazed sites (in the east and in the north: Sites 2 and 3), the foliar nitrogen levels dropped significantly following the input of sugar—that is, these forest limit sites showed a lack of resistance to the disturbance. The foliar nitrogen levels took a significant step down at the eastern grazed forest (Site 2) following the sugar disturbance—and dropped even lower at the northern grazed site (Site 3). In fact, the pattern was once again as clear as a staircase—only this time, nitrogen levels were going *down*.

Here was the answer to my question, *Does a woodland with a history of higher levels of disturbance have a more difficult time responding to additional stress than one with a lesser history of disturbance?* Yes. The woodlands carry those stress loads—memories of the stresses, so to speak—and this affects their ability to handle new stresses that come their way.

Maybe it was a coincidence that the highest levels of foliar nitrogen were at Site 1 (with the lowest stress levels) and the lowest levels of foliar nitrogen were at the disturbed "sugar plots" at Site 3 (with highest cumulative stress levels). But I don't think so. Here is why. More sheep equals more trampling. More trampling means less moss. Less moss means warmer soil temperatures (thick layers of moss keep soil cooler). Warmer soil temperatures mean increased microbial activity. Larger microbial populations means less nitrogen for birch trees.

As a picture of these feedback loops began to emerge, I remembered the metaphorical story of the flapping butterfly wings that trigger a series of reactions, ending in a wild storm. Taking such a chain reaction further at my study sites, I knew that fewer (or no) plants means no soil. No soil essentially means no ecosystem. While you might think that *no ecosystem* means *no sheep*, I saw—and more often than I would have thought—sheep picking their way across a gray landscape of rocky glacial till with hardly a blade of grass in sight, and certainly none dripping with butter.

At this point in the story, it sounds like sheep are, for the most part, nothing but bad news for the birches. Moreover, the cards are stacked against the birches not only because sheep are eating them but also because sheep give microbes a leg up in terms of competing for the available soil nitrogen. However, if sheep are nothing but bad news, one piece of the puzzle doesn't fit.

At the eastern protected site (Site 1), the moss layers were so thick that you could fling yourself backward into the moss (a moss "trust fall," so to speak) and land in a moss bed comfortable enough for even the pickiest of elves. A sturdy fence had excluded sheep at this woodland for a century, and the human footprint was similarly light. Is this site the most pristine? The most wild? I might be tempted to say yes, except for this: at this woodland, the altitudinal location of the forest limit was the lowest among the three sites.

The forest limit of the northern woodland (Site 3)—with plenty of sheep, the harsher climate, and a shorter growing season—definitely looked scrappier, lacking that lush layer of moss. This was not a good place for a moss trust-fall exercise—not a surprise, given that sheep trampling is not conducive to moss growth. The surprise was that this forest limit was located at *highest* altitude among my three sites. Despite the stresses present at this site, this woodland had managed to climb higher up the mountain than either of the other two sites. This struck me as an impressive feat. Even though I did not understand it yet, there was a consistent pattern: the forest limit at the eastern grazed forest (Site 2, with the middle of the road stress of the three sites) had the middle of the road scrappiness (and moss layers), along with a middle of the road altitude.

How could I solve this puzzle? The northern grazed site (Site 3) with the most stress (historical and current) had trees growing at the highest altitudes on the mountainside. With all that stress—at the higher altitudes and with the lower foliar nitrogen levels—one might wonder how these birches are surviving at all. But there they are. True enough, at these higher stress levels, the birch trees are closer to their ecological breaking point. Add just the right amount of stress (especially in the form of competition for nitrogen) and the birches at this northern site would reach a threshold where they could no longer function. In contrast, the eastern protected woodland (Site 1) was buffered. Higher levels of foliar nitrogen left the trees some wiggle room for taking on additional stress; it was more of a birch safety zone. That being said, the protected woodland in the east did not extend up the mountain nearly as far.

At the grazed sites, perhaps the warmer soil temperatures allowed for expansion of the birch woodland into higher altitudes. While the warmer soils may have allowed the birch to exist at higher altitudes, the trees at the grazed sites are also at a higher risk for nitrogen competition (from microbes enjoying the warmer soils) and grazing (from the aforementioned sheep). In other words, the birches at grazed tree lines exist higher up on the mountainside, but at the same time, they live closer to their edge. While this may not be the safest route for the birches, it is perhaps worth the risk, because the upside is pretty big: the chance at life.

It sounds familiar. Given the choice, I would rather be on the edge of human experience, certainly on the edge of human knowledge, and even tolerate the edge of emotional comfort—if it meant life. And does not history (our own and others') show that experiences on the edge can offer important insights into both what it means to be human and what it means to be one human in particular? For me, "living on the edge" is part of the daring—and the learning—that is central to the evolution of life.

There are many expressions of Iceland's wildness, and all these expressions depend on the presence or absence of sheep. Perhaps the most common depiction of the Icelandic wild involves Iceland's gray moonscapes, with sheep—and not trees. However, these starkly beautiful landscapes have crossed over an ecological threshold beyond which it is very hard to return. These landscapes are wild and wooly, but if you do not know how they came to be as they are, you may not be able to put your finger on the sadness that you might sense in the haunting gray vistas.

One could argue that the lush, protected woodlands are Iceland's most wild places, despite the fact that they are enclosed by human-made fences. These sheepless woodlands offer wild, green memories, seemingly borrowed from the time of the Vikings and carried into the present day by their human—and elf—protectors. On the other hand, in some places, Icelanders ask the Icelandic Forest Service *not* to plant more trees. The chief of the Icelandic Forest Service, Þröstur Eysteinsson, told me that in such cases he hears the complaint that trees will "ruin the view." "They are optimists,"

Eysteinsson retorts, because it is of course no small task to restore a whole forest ecosystem anywhere, much less in such a harsh climate.

If I were I to show you what I believe to be the wildest places in Iceland, however, I would take you to the forest limit, in a birch woodland populated with a good number of sheep and enough moss to satisfy the average elf. Mind you, this place would not have too many sheep and not too many soil microbes for that matter. I would take you to a place where birches breathe life into a landscape shared with sheep and their people, a place where the story told by both the sagas and the landscape itself is a story of life taking a chance — on the edge.

## Note

1. As noted by Þröstur Eysteinsson, chief of the Icelandic Forest Service (and one of my dissertation committee members), "Icelanders do not have the concept of the small winged creatures that English speakers call fairies." As for the elves, Þröstur says, "Our elves are human size and look like people, only better looking and better dressed."

## Reference

Anonymous (thirteenth century). *Landnámabók (Book of Settlement)*. In *Íslendingasögur (Sagas of the Icelanders)*, vol. 1, edited by Guðni Jónsson. Reykjavik: Íslendingasagnaútgáfan, 1946.

# 22

## THE STORY ISN'T OVER

*Julianne Lutz Warren*

The poet Rainer Maria Rilke was in his late twenties when he moved to Paris. The rattling nineteenth-century city tormented his raw nerves as he struggled with the meaning of "some unspeakable confusion . . . called life" (qtd. in Banville 2013). As the story goes, the sculptor Auguste Rodin advised his young friend to visit the zoo, focus on one animal until he knew his every move and mood, and then write about it. Rilke chose a panther. Studying this animal oriented the poet in the world as "a real person among real things," helping him reconcile with his change-making time and place (Banville 2013; Kentridge 2014). Rilke's now well-known "The Panther at the Jardin des Plantes" helps readers feel a genuine experience shared across species. The poem's three stanzas also are a helpful guide to contemplating what it means to live in a cage, which is critical when it comes to appreciating what may be both human and wild in our unfolding age.

> ***Stanza I: Beyond us the world exists no more***
> Ceaselessly the bars and rails keep passing
> Til his gaze, from weariness, lets all things go, for
> it seems to him the world consists of bars and
> railings, and beyond them the world exists no more.

The *Anthropocene*—a name first suggested by scientists to signify a time of unknown length characterized by human domination of the world—is gaining wider currency. Living amid it, we may feel that, with the rest of life, our whole humanity has been imprisoned by members of our own species with no possibility of escape. Like the panther who loses the world by seeing only bars and railings, we may become weary of gazing at the bounds of Human Empire—a steeled system run on the principles of profit over generosity, efficiency over healthy beauty and emotion, and power over dignity. The unintended consequences of this system now enclose every being on Earth from land to sea and bedrock to atmosphere, though not yet to the stars (though we know the stars are in us). In such a state, what, if anything, could *wild* mean?

* * *

In 1989, journalist Bill McKibben's best-selling book *The End of Nature* reported the increasingly perceptible transformation of the whole world by human beings. Of course, all living beings, just by breathing, alter things. But we humans now could see that we had *so* altered Earth, McKibben pressed, that we could bound its slowly spiraling time with a thin line marking where a "Before" ended and an "After" began. Indeed, crossing this threshold, the planet is now so different, he argued later, that it deserves a new name. He proposed *Eaarth* (2010).

To be clear, in suggesting a new name for our planet and in announcing the end of its nature, McKibben was declaring the cessation not of green mountains, meadow voles, wind, and rain but of *nature* as the *idea* of "a separate and wild province, the world apart from man to which he adapted, under whose rules he was born and died" (1989, 41). It was this *idea* of nature—as something wild out there beyond human reach and of which we were subjects—that had expired, he claimed.

As McKibben, in the 1980s, sat by an Adirondack Mountains waterfall, he knew that water, rock, and gravity were still water, rock, and gravity. But he also recognized that the reverberating consequences of Human Empire's activities now shaped the context of their presence. These activities included crossing seas to carry one land's fruits into another and severing them into "commodities" in exchange for a growing population of fewer wealthy and more desperate people; burning forests and bedrock-buried fossil hydrocarbons for expanding industry and speedy travel; plowing up long-coevolved intricacies of unique, diverse beings and upsetting their self-renewing relationships with soil; and replacing Earth's self-composting gardens with genetically engineered monocultures of irrigation-requiring, chemically amended crop regimes. And now McKibben could feel the consequences as pattering raindrops from a smoke-and-ashes-smoggy-greenhouse-gas-ridden sky onto craggy mountains and Earth-quaking plains, rippling into flowing currents downstream and outward to the shores of a future too distant to apprehend.

Dominating human activities now are reversing billions of years of net self-augmenting trends in an unfolding, global, life-unmaking event. Worldwide soil fertility is slipping away faster—much faster—than it is building up (Montgomery 2010), carrying toxins with it. The world-of-life is losing species faster—much faster—than new ones are evolving, heading into an unprecedentedly precipitous mass extinction event (Ceballos et al. 2015). Earth's atmosphere, in record time, has gathered such a thickening climate-warming blanket of greenhouse gases as to make it dramatically unlike any to which current life is adapted (Pachauri et al. 2014). The effects of climate warming are reinforcing the ecological destruction and energetic disordering that brought it about—effects that include high and low precipitation extremes; lashed-up wind gusts; melting ice and rising and souring seas; and erratically revolving seasons, desynchronizing serial dependencies, stressing uncertainty. A predictable pattern of two harvests a year in eastern Uganda has been undone, for instance (Okollet 2009). And now one year brings ruining floods swarming with malaria-bearing mosquitoes, while the next brings droughts with heat intense enough to wilt plantains, sweet potatoes, corn, and coffee, while drying up wells, livelihoods, and prospects for future generations.

Yes, the Adirondack waterfall is still a waterfall. As keen-witted humans connect the dots, however, we perceive that its rocks and water (if not unrelenting gravity)—like the bloodrooted soils of its banks eroding into the distant sea, which is inhaling fossil hydrocarbon fumes from air into its brew of melted ice, dissolving coral reefs, and teeming plastics—no longer exist independently of us.

As scientists continue deliberations over a precise geological marker for the start date of the Anthropocene (Lewis and Maslin 2015), we may, meanwhile, more generally acknowledge a time of profound changes on Earth with causes rooted in human culture, which itself shows signs of changing as a result. Understood as a cultural phenomenon, the Anthropocene begins roughly in the nineteenth century (Marsh 1869, 8) with a shift in shared ideas about human responsibility for global-scale influence. It is a shift that includes a complex, unwanted accumulation of consequences of human activities springing from the reductionist values of human imperialism—consequences cascading into a future of unknown length in uncertain ways. People who have crossed this line of altered awareness may appreciate the beauty, if not the health, of a real waterfall, the sky and sea, and of all Earth. With a rising consciousness that the world no longer exists beyond the expanded confines of Human Empire's influence, however, we may no longer experience the world as "other" and, paradoxically, also not feel alone. That is, we may feel, as McKibben puts it, both "lonely" and—as unable to escape the dominating presence of our own kind—"crowded, without privacy" (1989, 76).

In Manhattan, as I walk through Central Park, I listen for white-throated sparrows to sing come fall, having arrived from where they bred in the high north (Warren 2013). Their bright notes please me through winters of slushy sidewalks and short, cloudy days. As it seems to Rilke's panther that the world "consists of bars and railings," similarly, however, I can no longer hear the birds' voices apart from imposing human factors to which I contribute. That is, I can no longer hear the songs apart from the loud drone of the jet planes I fly in, the hum of my oil-heated radiator switching on, the whistle of my gas stove boiling tea-kettle water, and the words I read in an Exxon Mobil report (2015)—it is "possible," it states, but "difficult to envision" that governments will choose the path to net-zero carbon that would strand the company's billions of dollars in assets—blended with the cracking trees in spinning gusts and the emergency sirens accompanying the full-moon sea-surge flooding of 2012's record-smashing Superstorm Sandy. Nor, however, can I separate all this from more than four-hundred-thousand people, including me, whom McKibben's activist work with 350.org helped self-organize, moving down Central Park West two years later, chanting: "I hear the voice of the people singing / Exxon your kingdom must come down."

\* \* \*

Humans are responsible for profound global changes. Irrepressibly intensifying winds and lashing rains, gravity slipping soil, heat wilting crops, opportunistic "pests," and tidal upwellings demonstrate, on the other hand, that though empire tried to control it, nature has not become well trained. The sparrows' singing and the marching people's refrain likewise indicate that being entrapped is not the same thing as being tamed. There are uprisings of the Anthropocene that indicate some emergent reordering force that is deep, encompassing, and wild.

***Stanza II: Like a dance of strength around a centre***
Supple, strong, elastic is his pacing
And its circle much too narrow for a leap
like a dance of strength around a centre
where a mighty will was put to sleep.

The people vitally march and chant, but to bring down the stubborn kingdom fueled by oil and gas—while entangled with it—still strikes most as an overwhelmingly confusing task. An exercise of dreaming might help one better understand such transformational complexity: From a bird's eye perspective, we watch our own wills—like the panther's feeling no hope of escape—quench faith and desire, shut their eyes, and sleep. But with half-conscious power, relentless, our urging bodies—with restless strength, un-resting, if not leaping—not only march but rise up to dance around their centers.

Those dancing may feel, with each contracting muscle, the heaviness of multitudes of humans and other self-willed forms consumed by human power—those deemed of use, like dark-skinned Africans, wide-girthed hemlocks, energetic fossil algae and fecund prairie soil, oil-brim whales, tasty-fleshed auks, and many bison. As each muscle relaxes, we may feel the lightness of those of no foreseen service thrown out, disarrayed, crushed, and despised for being odd, too little or too much—like all sorts of mussels, darters, and tiny rockwrens, monarch butterflies, carbon dioxide, wolves and big cats, the Havasupai, and the girl working at the checkout—that is, we feel the lightness of unmattering.

As we flex, we feel something we may have forgotten but long have known within our flesh and bones: Though currently entangled with an imposing human system, each being's intimate intermingling with others is far more deeply essential in the enduring world-of-life. As we consciously reflect on this—and on the moon, sun, and universe of stars—we see that the end of the idea of nature apart from humans is the end, too, of the idea of humans apart from nature. This awareness begets mourning over lost parts of Earth that we belong to. It also upsets the rational-imperialist worldview, troubling the meanings of what many of us once thought we knew.

\* \* \*

"From a global perspective," writes American novelist Jonathan Franzen, "it can seem that the future holds not only my own death but a second, larger death of the familiar world" (2015)—a world that we humans signify with words. We English-speakers of the civilizing West once said *wild*, *bird*, *flower*, *soil*, *city*, *ocean*, *wilderness*, and *hope* and thought we understood what we meant. As empire builders, we also said *farm* and *road* and *factory* not long before we said *atmosphere*. And then *dumpage* bought grounds for ruination while those privileged fought to stay that way even as they proclaimed slaves' *emancipation* (OED 2015). We apperceived each word according to our culture's defining blueprints as an indomitable body with roots or feathers, a place of universities and sewers, big lands of fishes and bears, fruiting trees and safety, a tract to cultivate, a traveling way between places, smokestacked buildings for making things, the air around the earth, a costly site for piling excrement and mine tailings mixed with plastic forks, and release from captivity. But in our Anthropocene transition, we can see lines fading and redrawn, raising many questions about weaves of interrelations.

We can see the whole Earth as viewed from space. The wakes of ships and contrails of jet planes arc back and forth from place to place, carrying raw materials and their products, packages of everything, moving around all kinds of people and other species—some intentionally, like Peruvian guano to Great Britain's soils and Indian minas to New Zealand; some by accident, like tiny ticks on minas and the hemlock woolly adelgid from Japan and garlic mustard from Europe delivered to the United States—rapidly jumbling all sorts of materials and beings. While unseen currents, in mere seconds, transmit matter-altering ideas between neighbors next door or across a hemisphere—like how to dress as fashion conscious, bake a chocolate cake, and drill or prevent drilling an oil well. Where is there not a traveling way? Are there boundaries between oceans? Can we divide flowers, birds, and soils?

As from a distance, we keep watch as Earth slowly turns from sun toward moon, growing dark. We see ground constellations of lights flick on. These tend to be brighter not only where human populations are most dense but also in places of greatest wealth (NASA 2000; Davies, Lluberas, and Shorrocks 2012). The United States—one of the world's most brightly lit, populous, and wealthiest nations—has dumped into the whole Earth's atmosphere more than a quarter of the anthropogenic carbon dioxide of the past two hundred fifty years (Hansen 2007, fig. 27). Within the past decade, China has overtaken all other countries in carbon emissions and, though agreeing to mitigate (The White House 2014), has an expanding energetically consumptive population. India and Africa—that is, Uganda plus all its other countries put together—have each contributed barely 3 percent of rising greenhouse gases, and human numbers on their continents are among those mounting fastest. Worldwide now there are more than seven billion of our species and counting. We all must eat, drink, warm ourselves, wear and build things, and travel to survive. Many want more, to prosper, quickly spinning outward in rutting traces. Where does a factory end and atmosphere begin? Can we clearly mark out nations or cities, farms, and wilderness?

In response to human expansion, Earth whirls in now on everyone, heedless of responsibility, rights, contracts, maps, and fences. And, so, therefore, as aroused by privileged hands—unjustly—most harshly delivers alarming, undesired, rippling consequences to those living in the least well-lit places (Pope Francis 2015). Can we really set off dumpage from human beings? What might hope represent today in a world of diminishing safety and for future generations of all life?

It is July's early dawn in near-Arctic-interior Alaska. I am running through hills under pale-gray skies, half asleep. As I come down a bend, my view suddenly opens onto a vast boreal forest reaching the horizon. The low-hanging sun appears so brilliantly red that, without willing it, I utter a loud "Oh!" and stop dead in my tracks. I hear a chickadee sing just then. I breathe in. My lungs feel heavy with yesterday's Fairbanks *Daily News-Miner* headline: "Wildfire season of 2015 may soon be state's worst." It is the not-too-distant veil of smoke rising from millions of acres of heat-stressed, lightning-struck trees that is coloring the solar rays. I breathe out. I feel the lightness of the wildfire carbon invisibly blanketing the sky and of the dark ashes that will fall back to Earth and renew the thin soil. On the warming planet, authentic hope—defined as this place—is likely to change, almost before our eyes, into bluebunch wheatgrass, sage brush, and buffalo berries or aspens rather than mossy white spruce and those that go with it.

* * *

As we lose the familiar world, in bodies and words, there is much to lament. As we begin reapperceiving events in relation to the obdurate-entwinement of humans and the rest of nature, genuine tales of weary confinement unfold. But we may also sing, if not of escape, of mutual emancipation by something sailing through the openings between rails of a broken system—something that is still recognizable, strong and supple—a recomposing wild with which we may participate.

### *Stanza III: The pupils' curtain rises silently*
Yet from time to time, the pupils' curtain
rises silently. An image enters, flies through

the limbs' intensive stillness
until, entering the very heart, it dies.

As if his will and strength are reconciled, the panther appears to pause now in his caged dance and, though quiet, to come mostly awake. His eyes blink as do ours, each glimpsing the other as through opposite sides of iron bars. Images of us enter the big cat, while the poet's words also fly pictures of him into us. We discover in this exchange, again, some essence that we share, which is far older than us, in our "very heart." Though those chambers are hard to reach, they are alluring and we long for entrance. That mutual yearning sparks a possibility: that hearts fierce enough to kill each apparition of the other could also muster sufficient ardency to help each other leap, with grace, into regenerative stories.

* * *

In Wellsian time-machine fashion, imagine landing near the top of the world, in the northeastern reaches of what is now called Siberia. We've zoomed backward into the late Pleistocene some thirty-two thousand years ago. Earth's climate is slowly fluctuating on the threshold of another long cooling. It is a few thousands of years before the first human beings actually do arrive to this land that drains chilling waters into the Kolyma River — a wide rush of meltwater and rain — into the Arctic Ocean. The river's banks stretch out into steppe-tundra. In brief spring and summer, the vast place blooms with flowering plants. Their green leaves suck up the midnight sun's light, make sugar, grow, and hoard their sweets in fruits and seeds to grow future roots and leaves. This severe place is full, too, of plant eaters, including mammoths, wooly rhinos, musk oxen, reindeer, and ground squirrels. The squirrels carry multitudes of the faded flowers' nutrient-rich parts underground to their fur-lined burrows, injuring many of the mature germs, likely for better-keeping storage (Gyulai et al. 2011). There are carnivores, too, of course — wolves, wolverines, bears, foxes, and cave lions. Who knows who eats which of the squirrels leaving their caches behind, while those fed, uneaten, and uncached go on with living overhead? The deep chambers seal with windblown earth. The buried deposits wait in the dark, frozen more than thirty meters belowground in permafrost for hosts of generations.

Until now.

The Kolyma and its wide flanks of land have weathered impermanence over long ages with changing climates and shifts in plants and animals, including extinctions—as of mammoths, rhinos, and cave lions—and new arrivals, including human beings. There are also constant lines of things. The place still blooms with many of the Pleistocene's bequest of flowers—for instance, generations of *Artemisia*, *Phlox*, and *Silene* species—that carry on feeding herbivores, including ground squirrels. The squirrels no longer feed lions but do help sate foxes, wolves, and other meat-eaters. Joseph Stalin's gulag prisoners have come and gone. Others have lived on. The Yukaghir hunters, the first people—that is, those who have endured longest here—have dwindled to mere hundreds. Most recently, lured by sparkling cities, some of them have moved away. An old remaining hunter sings:

> My land, winter here is very cold
> Yet plants grow despite the cold
> My life is nearing its end
> but my land has remained the same as in the time of
> my youth
> It is good and generous
> it gives us everything
> and has not forgotten us yet. (Lecomte, c. 1994)

Today's changes are unprecedentedly rapid, however, making it difficult to find footing on what stays constant. With Anthropocene climate warming, this good land—frozen for so many thousands of years—no longer remains so much the same across even one human generation (Banerjee 2013). But alongside death, generosity seems to endure. As the Arctic warms at a rate nearly twice the global average, the erosion of the Kolyma's ice-filled banks has been advanced by melting permafrost, also sending alarming amounts of more carbon into the air. On the quieter side, the softening soil offered scientists the chance to find and excavate the ancient ground squirrel capsules—released, but as yet unthawed. Some are filled with hundreds of thousands of their hosts' seeds and fruits. Many of these are biologically dead. But others, *Princess Bride*-like, are "only *mostly* dead"—that is, as imagined in the film, they are *as if* dead without the proper magic.

If we dig into the history of science, we find a preoccupation with magic (Fara 2009, 101)—in other words, with learning how to flow with nature's enigmatic winds and waves into our desires. Magic's rituals concocting brews and chants went to our heads some time back, however, and some got to thinking that our journeys would be easier if we could build more certain paths. Fruits, like hearts, however, cannot long be held without inviting wonder at the blueprint-less mysteries from the past and future secrets they enfold.

So—as if bidden by a deep, complex urging—a group of scientists, curious, tried reaching into the squirrel-hidden stores of fossil plants to coax some of them to waking life. In their lab, just south of Moscow, the cotyledon of a sedge embryo enlarged but never left its seed to become leaves; radical cells of bearberry divided but did not form roots; a sourdock stopped just short of sending out its shoot, and a severed part of its cotyledon placed in agar found wherewithal to almost mend itself, formed calluses, but still aborted; as did, too, of special note, some radical bits of *Silene stenophylla*, a pale-flowering plant whose arctic legacy has gone on adapting and developing, remaining extant. Now, as if reaching back out, these ancestral shivers of thawing vitality forecasted future leaps.

The ancient plant-caching ground squirrels tended to leave unripe fruits, which were too immature to germinate, uninjured and intact. The ardent researchers carefully cut into some of the *Silene*'s paradoxically young fossil fruits and removed slips of their placentas. These, it turned out, had kept well with cold-protective and nourishing elements tailored for their undeveloped progeny. Taking cues from the plant, the scientists fed these tissues freshly concocted brews of salts and vitamins, splashed with coconut milk. In response, some of those tissues sent out shoots, then roots that took to soil, blossomed, and developed a new generation of fruits fecund with seeds.

Scientists, in their own words—voiced as human beings with both thoughts and feelings—were "amazed" and "excited." Indeed, plain people around the world seemed to gasp collectively. The news of Earth's now far-oldest known flowering plant, first published in a top scientific journal (Yashina et al. 2012), was picked up immediately by venues worldwide. Headlines read like that of *The Guardian*, which announced on February 21, 2012: "Russian Scientists Regenerate Ice Age Plant." The far-flung stories were accompanied by closeup portraits of a small, resilient beauty with five white heart-shaped petals and often repeated not only the word *regenerate* but also *revive* and *resurrect*.

What seems striking in these reports is not only the skill of the scientists helping the very old fruits bring forth their new flowers but how those fruits' flowering helped resuscitate widely in humanity—beside vital panther fierceness—ardent tenderness, joyful camaraderie, and shared relief at some fresh idea of the world-of-life not yet ending. As a recent *National Geographic* essay put it, "The story isn't over" (Krulwich 2015).

*\*  \*  \**

This unended story is of the unmaking and remaking of sky-to-root-tooth-to-fruit-heart-to-eye-to-eye-to-sky wild. And, given grace to endure, it may include the ongoing evolution of humanity, turning through the *Anthropocene* into supple dancers reflecting our belonging with a multi-centered *Eaarth*—our belonging with a self-augmenting world-of-life that has, with our participation, quickly resumed billions-of-years-old movements of unfurling diversity and complexity, empathy with community, and freshening cultures of expanding consciousness reconciling knowledge and generous morality into an unknown future. For expressing such a new *idea* of an already budding offshoot of our own species, I propose a new name—*Homo generativus*.[1]

## Coda: Ritual for Wild, Now

Five, four, three, two, one . . . *silence*

In a symbolic gesture, the four-hundred-thousand voices gathered for the People's Climate March in New York City went dead quiet in a moment for remembering. Each person in the crowd was encouraged to pay her own private honor to those already consumed, displaced, caged, and disarrayed as a consequence of human empire.

*Hear absences here.*

Memory with lament postures us toward humility and away from overconfidence in ourselves. This orientation lifts up "a new kind of people" (Leopold 1944; Warren 2006, 264) to appreciate the multiple possible meanings of relationships—like that between human beings and *Silene*—without continuing an other-dominating legacy of past mistakes.

. . .
ROAR < < < < < < < <

After the silence, a swelling wave of blended shouts, chants and songs, banging pots and pans, and brass bands began downtown and moved up the street. The crowd cheered for ongoing life—for Earth's remaining diversity of small and large, of useful and seemingly odd and useless human and other beings, for being themselves, who, sharing a persistent nature to self-organize by complementing one another in their common cause, must swamp human empire.

*Now, find the beat.*

Participation in a chorus of desires entices us toward possibility and away from being overwhelmed by ourselves and overwhelming others. This membership quickens a new people to listen, smell, touch, taste, and see—to know—each other as unfathomable "very hearts," longing beings, and to perform measures of skillful love, generatively.

## Note

1. *Generativus*—with a blended meaning from works of natural historian Charles Darwin, ecological conservationist Aldo Leopold, psychologist Erik Erikson, mathematician and philosopher Brian Swimme, world religion and ecology scholars John Grim and Mary Evelyn Tucker, and paleoanthropologist Rick Potts.

## References

Banjeree, Subhankar. "Let Us Now Sing about the Warmed Earth." *Huffington Post*, July 28, 2013. http://www.huffingtonpost.com/subhankar-banerjee/let-us-now-sing-about-the_b_3667581.html.

Banville, John. "Study the Panther!" *New York Review of Books*, January 10, 2013. http://www.nybooks.com/articles/archives/2013/jan/10/study-panther/.

Ceballos, Gerardo, Paul R. Ehrlich, Anthony D. Barnosky, Andrés García, Robert M. Pringle, and Todd M. Palmer. "Accelerated Modern Human-Induced Species Losses: Entering the Sixth Mass Extinction." http://advances.sciencemag.org/content/advances/1/5/e1400253.full.pdf.

Davies, James, Rodrigo Lluberas, and Anthony F. Shorrocks. "OECD World Forum New Delhi: Measuring the Global Distribution of Wealth." October 17, 2012. http://www.oecd.org/site/worldforumindia/Davies.pdf.

Exxon Mobil. "Energy and Carbon: Managing the Risks" (Report, 7, 11). http://cdn.exxonmobil.com/~/media/global/files/energy-and-environment/report---energy-and-carbon---managing-the-risks.pdf.

Fara, Patricia. *Science: A Four Thousand Year History*. New York: Oxford University Press, 2009.

Franzen, Jonathan. "Carbon Capture." *New Yorker*, April 6, 2015. http://www.newyorker.com/magazine/2015/04/06/carbon-capture.

Gyulai, Gábor, Lilja Murenyetz, Zsigmond G. Gyulai, Viacheslav L. Stakhov, Svetlana G. Yashina, and Stanislav V. Gubin. "Morphogenetics of *Silene stenophylla* Seeds from Permafrost of the Late Pleistocene (32–28,000 BP)." In *Plant Archaeogenetics*, edited by Gabor Gyulai, 3–9. Hauppauge, NY: Nova Science, 2011.

Hansen, James. "Figure 27." http://www.columbia.edu/~mhsl19/UpdatedFigures/. Data are sourced from Carbon Dioxide Information Analysis Center, Oak Ridge National Laboratory, and British Petroleum.

Kentridge, William. *Six Drawing Lessons*. Cambridge, MA: Harvard University Press, 2014.

Krulwich, Robert. "Seeds that Deified Romans, Pirates, and Nazis." *National Geographic*, July 28, 2015. http://phenomena.nationalgeographic.com/2015/07/28/seeds-that-defied-romans-pirates-and-nazis.

Lecomte, Henri. *Sibérie 3: Kolyma: chants de nature et d'animauz (Chuckch, Even, Jukaghir)*. Buda, Musique du Monde/Music from the world, 92566-2, c. 1994.

Leopold, Aldo. Letter to Douglas Wade, October 23, 1944. University of Wisconsin, Leopold Papers 10–8, I.

Lewis, Simon L., and Mark A. Maslin. "Defining the Anthropocene." *Nature* 519 (March 12, 2015): 171–80.

Marsh, George Perkins. *Man and Nature or, Physical Geography as Modified by Human Action*. New York: Scribner, 1869.

McKibben, Bill. *Eaarth: Making Life on a Tough New Planet*. New York: Times Books, 2010.

———. *The End of Nature*. New York: Random House, 1989.

Montgomery, Richard. "Is Agriculture Eroding Civilization's Foundation?" *GSA Today* 117 (2010): 4–9.

Okollet, Constance. "Climate Change Is Killing Our People." *Guardian*, September 23, 2009. https://www.theguardian.com/commentisfree/cifamerica/2009/sep/22/united-nations-climate-change-uganda.

Oxford English Dictionary Online. "Hope." http://www.oed.com/view/Entry/88370?rskey=AdE2fY&result=1.

Pachauri, R. K., et al. "Climate Change 2014: Synthesis Report: Summary for Policymakers." IPCC Report. https://www.ipcc.ch/pdf/assessment-report/ar5/syr/AR5_SYR_FINAL_SPM.pdf.

Pope Francis. *Praise Be to You—Laudato Si': On Care for Our Common Home*. San Francisco: Ignatius Press, 2015.

Warren, Julianne. *Aldo Leopold's Odyssey*. Washington, DC: Island Press, 2006.

———. "Winter Song." *City Creatures* (blog), Center for Humans and Nature, February 25, 2013. http://www.humansandnature.org/blog/winter-song.

Weier, John. "Bright Lights Big City." *NASA Earth Observatory*, October 19, 2000. http://earthobservatory.nasa.gov/Features/Lights.

White House. "US-China Joint Agreement on Climate Change" (press release), November 12, 2014. https://www.whitehouse.gov/the-press-office/2014/11/11/us-china-joint-announcement-climate-change.

Yashina, Svetalana, Stanislav Gubin, Stanislav Maksimovichb, Alexandra Yashina, Edith Gakhovaa, and David Gilichinsky. "Regeneration of Whole Fertile Plants from 30,000-y-old Fruit Tissue Buried in Siberian Permafrost." *Proceedings of the National Academy of Sciences* 109 (March 6, 2012): 4008–13.

# 23

## CULTIVATING THE WILD

*Vandana Shiva*

> Ganga, whose waves in Swarga flow,
> Is daughter of the Lord of Snow.
> Win Shiva, that his aid be lent,
> To hold her in her mid-descent.
> For earth alone will never bear
> These torrents traveled from the upper air.[1]

Treks to the source of the Ganges are among my fondest memories of childhood. At an altitude of 10,500 feet stands the Gangotri, where a temple is dedicated to Mother Ganga, who is worshipped as both a sacred river and a goddess. A few steps from the Ganga temple is the Bhagirath Shila, a stone upon which King Bhagirath supposedly meditated to bring the Ganges to the earth. The shrine opens every year on Akshaya Tritiye, which falls during the last week of April or the first week of May. On this day, farmers prepare to plant their new seeds. The Ganga temple closes on the day of the Deepavali, the festival of lights, and the shrine of the goddess Ganga is then taken to Haridwar, Prayag, and Varanasi.

The story of the descent of the Ganges is an ecological story. The above hymn is a tale of the hydrological problems associated with the descent of a mighty river like the Ganges. H. C. Reiger, the eminent Himalayan ecologist, described the material rationality of the hymn in the following words: "In the scriptures a realization is there that if all the waters which descend upon the mountain were to beat down upon the naked earth, then earth would never bear the torrents... In Shiva's hair we have a very well known physical device, which breaks the force of the water coming down... the vegetation of the mountains." The Ganges is not just a giver of peace after death—she is a source of prosperity in life. She is the source of energy—ecological, cosmological, economic, social, and cultural energy—for the people of India.

In the dominant paradigm, "energy" refers to oil and coal, which are mined from the earth, shipped thousands of miles, and transformed into electricity to light up neon signs or into fuel to run SUVs.[2] It is fruitful to remember that energy has other meanings and other forms. From Shakti, the generative force of the universe, to the sun that powers our lives, to the water that comes to us as bountiful rain or a flood or a tsunami or trickling and then roaring from Gangotri to fuel a living universe, to the air and the wind that move the clouds and create the climate. Energy is not just oil and gas. Energy is an all-pervasive element of life.

The broader our paradigm of energy, the wider our choices as human beings. Fossil fuels have fossilized our imagination, our potential, our creativity, our wildness. We need to break free from this fossilization to choose life-enhancing pathways for our selves, our species, and the planet.

The mechanistic paradigm has robbed us of our freedom and creativities. It has replaced living energy with fossil fuels; the wealth created by nature and people with capital; the freedom of citizens and communities with the coercive power of the corporate state that imposes the rule of capital in every dimension of our life—the thoughts we think, the food we eat, the settlements we shape.

In this life-threatening period of globalization and climate crisis, we need to unleash our hidden energies to make a transition to a post–fossil fuel economy. To do so, we need to reinvent democracy. A renewable-energy economy will only be built through the renewable energy of free and self-organized citizens and communities. The transition beyond oil is not merely a technological transition—it is above all a political transition in which we stop being passive and become active agents of transformation by recognizing that we have the capacity, the energy, and the creativity to make the change.

Life is based on the self-organizing energies of the universe, from cells to Gaia, from communities to countries. We as living systems are networks of chemical and energetic flow and transformation. Thus life is energy—not fossil fuel energy, but *living* energy.

The editors have asked me if this kind of renewal, if this kind of self-organizing energy, speaks to my view of "wildness." Cartesian duality and the colonizing worldview left us with a legacy that defined the wild as free of humans and the cultivated as free of nature. On the one hand, this led to the extermination of all that was considered "wild" in the colonized and settled zones—from the bison to the native people. It created an industrial agricultural model based on designing violent tools to exterminate biodiversity and everything alive—from the chemical fertilizers that kill the soil organisms that feed us, to the pesticides that kill the friendly insects (including bees and butterflies that produce one-third of our food), to the herbicides that kill plants that are food for humans and other species. The destruction of the milkweed by spraying Roundup is now recognized as contributing to the decimation of the monarch butterfly. On the other hand, it created a paradigm of wildlife conservation and national parks based on driving out the people who had conserved those places and their biodiversity. India is riddled with conflicts between people and parks because of this artificial legacy of the human-less wild, which becomes an inhuman construction of the wild. For the last four decades, I have worked to bridge this artificial divide by cultivating the wild.

Life is self-organized *autopoiesis*. For me, that is the definition of being wild. We save and conserve seeds because in their self-organization lies their wildness. We do participatory breeding as the cocreative coevolution between intelligent seeds and intelligent peasants. When we cultivate biodiversity as *Navdanya* (nine seeds) or *baranaja* (twelve seeds), we are cultivating the wild, not just in terms of the self-organized cooperation among different plants, but also between the plants and soil organisms, between plants and pollinators, and among the community of insects, including those that control pests without pesticides and GMOs.

By trusting the wild, we grow more food than what is produced through the use of violent tools and the violent attempt to "feed the world" by attempting to kill the wild. We have 350 percent more nitrogen in our soils than farms that use synthetic nitrogen fertilizers. We have six times more pollinators than in the forest next door.

Food grown through cultivating the wild is also healthier. The toxics like Roundup and Roundup Ready GMOs are not just killing the wild outside;

they are killing the wildness we need in our gut biome for our health. It is claimed that Roundup and Roundup Ready crops are safe for humans because (unlike the metabolic routes of bacteria, fungi, algae, and plants) humans do not have the shikimate pathway. This is outright violence against science. Ninety percent of the genetic information in our body is not human but bacterial. Out of the six hundred trillion cells in our body, only six trillion are human; the rest are bacterial. And bacteria have the shikimate pathway. The bacteria in our gut are being killed by Roundup, which is leading to serious disease epidemics, from increasing intestinal disorders to possible neurological problems such as increases in the occurrence of autism and Alzheimer's (Samsel and Seneff 2013). The soil, the gut, and our brain are one interconnected biome—violence to one part triggers violence in the entire interrelated system. Data from the US Centers for Disease Control and Prevention (http://www.cdc.gov/ncbddd/autism/data.html) show alarming trends related to the percentage of children born with some form of autism in the United States. It is not an intelligent species that destroys its own future because of a distorted and manipulated definition of science. As Einstein observed, "Two things are infinite: the universe and human stupidity . . . and I'm not so sure about the universe."

When I talk about the infinite creative energy of the universe, I am talking about Gaia's self-organizing energy, the creative human energy to work and to produce, to organize, and to transform. In India and around the world, this human energy has helped *cultivate* the self-organizing energy (whether a culture calls it Shakti or wildness) of the world. In particular, the creativity, innovation, and decision-making power of women (who still produce 80 percent of the world's food) has significantly driven the world's biodiversity. The majority of the eighty thousand plant species that humans have cultivated have emerged from the self-organizing, living energies of women. In other words, if we are going to redefine wildness, we have to simultaneously redefine humans as cocreators of wealth with nature. We both rely on and cocreate wildness when our living energies work with those of the earth.

Fourteen miles beyond Gangotri is Gaumukh, a glacier formed like the snout of a cow that gives rise to the Ganges.[3] The Gaumukh glacier, which is twenty-four kilometers in length and six to eight kilometers in width, is receding at a rate of five meters per year. The receding glacier of the Ganges, the lifeline for millions of people in the Gangetic plain, has serious consequences for the future of India. We need to generate and multiply the renewable energy of ecology and sharing, of solidarity and compassion, to counter the destructive energy of greed that is creating scarcity at every level—scarcity of work, scarcity of happiness, scarcity of security, scarcity of freedom, and even scarcity of the future. Either we can let the process of destruction, disintegration, and extermination continue unchallenged, or we can unleash our creative and wild energies to make systemic change and reclaim our future as a species, as part of the earth family. We can either keep sleepwalking to extinction or wake up to the potential of the planet and ourselves.[4]

## Notes

1. This quotation and the opening two paragraphs are adapted from my book *Water Wars*, 134–35.

2. The following five paragraphs are adapted from my book *Soil Not Oil*, 134–37.

3. The opening three sentences of this paragraph are from my book *Water Wars*, 135.

4. The final three sentences of this essay are adapted from my book *Soil Not Oil*, 137.

## References

Centers for Disease Control and Prevention. "Autism Spectrum Disorder (ASD): Data and Statistics." http://www.cdc.gov/ncbddd/autism/data.html.

Samsel, A., and S. Seneff. "Glyphosate's Suppression of Cytochrome P450 Enzymes and Amino Acid Biosynthesis by the Gut Microbiome: Pathways to Modern Diseases." *Entropy* 15 (2013): 1416–63.

Shiva, Vandana. *Soil Not Oil: Environmental Justice in an Age of Climate Crisis*. Cambridge, MA: South End Press, 2008.

———. *Water Wars: Privatization, Pollution, and Profit*. Cambridge, MA: South End Press, 2002.

# 24

EARTH ISLAND

*Prelude to a Eutopian History*

Wes Jackson

## *The Great Awakening*

Not long ago, our ancestors had a Great Awakening. It began small, but gained in size in the twenty-first century. What aroused the people on Earth Island? Global warming was the major concern then, and land use was the second-largest source of greenhouse gases, behind fossil fuel power plants and ahead of all transportation. Land was being degraded at a rate of thirty million acres per year. Our food source, the soil, was under siege. The world population, still growing in 2016, had tripled in the previous eighty years.

Our ancestors had a hard time getting a grip, primarily for two reasons: they were addicted to fossil fuels, and they continued to enact ideas of more distant ancestors, the children of the Enlightenment, who set their goal on "enlarging the bounds of human empire to the effecting of all things possible" (1857 [1620], 156)—a reductive approach to the world. Many of our ancestors' motivations were good. Like us, they wanted a world without hunger. A certain industrial heroism prevailed among them. Their dominant slogan was telling: "We must feed the world!"

Thankfully, other more modest people questioned, "Sure, the world must be fed, but then what if we fail to stop greenhouse gas accumulation, soil erosion, and depletion of fresh water?" These were the days of the Great Awakening, when few appreciated that soil is more important than oil and just as much of a nonrenewable resource. Millennium Ecosystem Assessment scientists concluded that agriculture was the number one threat to Earth Island's wild biodiversity. It was time to confront the *Problem of Agriculture* instead of only addressing problems *in* agriculture; it was time to tap the wildness, the diversity and creative complexity of the entire ecosphere, beginning with the land that feeds us; it was time to awaken a new and equally creative, diverse, complex, and equally wild human place within that ecosphere.

An increasing number of agricultural scientists and ecologists accurately diagnosed the negative consequences of grain production. Noting that natural systems mostly featured perennials and more or less constant ground cover, they called for *ecological intensification*. That is about where the agreement ended, for there were already two camps of agricultural scientists ready to address the problem.

The dominant camp was like most early twenty-first-century scientists. The generation before them—indeed, some of their major professors—were the agriculturists responsible for the twentieth century's Green Revolution, which, by industrial agriculture standards, was a great success. That is what the new generation sought to replicate and intensify with greater technological ingenuity. After all, grain yields had doubled, sometimes tripled. Almost everywhere into the twenty-first century, the Green Revolution was still hailed as a success—and in a limited sense, it was.

Obvious to all, but seldom mentioned, was the revolution's fossil fuel dependency, degraded soils, poisoned water, and greatly accelerated greenhouse gas emissions from agricultural land. In less than half a century, farms ecosphere-wide industrialized and increasingly depended on this extractive economy. Before the new varieties could respond, fertilizer, pesticides, and when necessary and possible, irrigation wells were required.

The Green Revolution success story had another shadowy side: the collateral damage to the small farmers who, for credit, mortgaged their farms to buy inputs. When the inevitable crop failure came, they lost their small farms, usually to larger farmers. Where was the more thoroughgoing critique of this to come from? Not from the scientists who drove the process forward.

The implicit and explicit goals of the Green Revolution shock us now, especially the implicit assumption that agriculture was not vitally linked to nature! With the Green Revolution, more than food was on the line. Economic development was also, and for good reason: two world wars and a global depression. Agriculture was to serve as an instrument for the advancement of industry and economic development. Adoption of the entire Green Revolution package was considered essential. Technology was considered neutral. If persuasion failed, compulsion was considered necessary. If the core problem was low productivity, then chemicals were relied upon as the solution. If more chemicals were needed, then more chemical plants were built. Soil degradation was factored in, but as a cost in the economist's input/output ledger. The gap between social and scientific cultures was largely ignored. Techniques of traditional farmers were regarded as more of an obstacle than a resource. Those in the developed world thought of themselves as the teachers. The poor were considered learners.

By the twenty-first century, agricultural scientists had heard the critique and were aware of the social and physical consequences. They wanted to do better next time. That is how "ecological intensification" became a mantra. But what did that mean? Well, more technological cleverness: precision planting, molecular tools, GMOs. Biological diversity was given a nod through rotations and cover crops. A combination of agriculture engineers, molecular biologists, geneticists, and agronomists all readied themselves to launch a second, but this time more benign, Green Revolution.

But a problem persisted. The second effort at a Green Revolution shared the same paradigm as the first. Given past successes with annual grains, technological cleverness still reigned. Radical thinking was difficult. The dominant camp still understood annual grains as the necessary "hardware," even though the "software" of how to grow them had proven limited. Grain fields need high nutrient retention. Agricultural fields need to accumulate organic matter and manage soil water. These goals are hard to meet once vegetation is cleared from the field, whether with the plow or with herbicides. Annual grain plants were not in the ground long enough for soil microbes and invertebrates to fully protect and enhance the soil quality. Soil erosion continued.

There were important voices crying in the wilderness. Professor Angus Wright, a historian of Latin America and the environment in the twentieth century, wrote books and articles that are still widely read today. Wright was a major diagnostician of the assumptions leading to the negative consequences of the Green Revolution. He was also a student of the movement of landless people. Social and environmental justice, he believed, went hand in hand. Others felt the same way. And so, beset by global threats, a different way of thinking emerged—a search for the agricultural virtues of wildness, a journey toward a more thorough intertwining of human and ecological health. Part of this search included the launch of a major effort in the second decade of the twenty-first century to domesticate more herbaceous perennials, to increase the variety of "new hardware."

This new effort made possible the new paradigm. A few scientists looked to nature's natural ecosystems for their standard—those perennial grain mixtures that had evolved over millions of years. That represented a better form of *ecological intensification*! Would it not be easier to achieve the goal that both camps agreed on if they were to mimic a prairie or grassland with perennial grain mixtures? Perennial hardware made it possible in one stroke. It took half a century. The paradigm for grain agriculture, which had existed for ten thousand years, was changed from conquering the wild to understanding it, working with it, even helping produce it.

An effort was launched to do a thorough inventory of herbaceous perennials and shrubs that produce hard seeds. Researchers also set themselves the task of analyzing what happened during domestication of annual grains and from there explored how it could be repeated with their wild perennial candidates. They were encouraged by the rapid response to selection of two wild species: intermediate wheatgrass and *Silphium*, a relative of the sunflower. The former became Kernza and the latter an important oil seed crop. Kernza, a relative of wheat and other grains, allowed plant breeders to transfer knowledge from those other grains. What is known as the $q$ *gene* was discovered in Kernza (a gene that is also present in wheat), which allowed for both shatter resistance and free threshing. New genetic techniques helped achieve breakthroughs in domesticating complex wild species. Domestication also required evaluating large numbers of plants and selecting the best to intermate. New computational power made that possible.

But there was more to the equation. It had long been known that any new crop requires more than breeding and genetics. Interdisciplinary teams featuring students of agronomy, plant pathology, soil science, food science, plant breeding, economics, and social justice were assembled.

Unlike agriculture scientists before them, the people on these teams were mindful of what Wendell Berry had to say as far back as 1974: "Few people, whose testimony would have mattered, have seen the connection between the modernization of agricultural techniques and disintegration of the culture and the communities of farming." As for those who would develop the crops and grow the food, the scientists knew, as Berry did, that "in the long run, quantity is inseparable from quality. To pursue quantity alone is to destroy those disciplines in the producers that are the only assurance of quantity. The preserver of abundance is excellence."

## How Did It Happen?

Well, it didn't just happen. Late in that period, a few established scientists had been watching, reading the papers of a few young scientists scattered here and there. These younger scientists wanted nothing less than to bring the processes of natural ecosystems to the ecosphere's grain fields. The elder scientists agreed that agriculture needed a fundamental course correction, but they frequently asked, "What makes you think acceptable yields can be attained out of such a diverse system?" Two reasons, the young scientists answered. First, it is well known that natural ecosystems tend to have greater net primary production than the annual monoculture systems managed by humans. So, to the extent that we can imitate wild structures and processes, we have data to support that we can be granted high net primary production at the ecosystem level. Second, at the individual plant level, perennials can have a longer growing season. Because annuals tend to accept their own pollen—the tightest form of inbreeding—the mutation load does not build up. It gets purged every generation. Perennials, on the other hand, tend to outcross, to not accept their own pollen. The mutation load accumulates. When closely related offspring are crossed, lethal or otherwise undesirable genes lead to aborted embryos or otherwise undesirable plants. Our ancestors lacked this knowledge—and the know-how for dealing with it.

Geneticists finally had the computational power and molecular tools they needed. They grew out tens of thousands of plants and purged the genetic load. They still made wide hybrids between wild perennials and their annual relatives, wheat, sorghum, sunflower, and rice. They began solving the oldest environmental problem: grain agriculture.

The new "hardware" won the geneticists new colleagues. These young ecologists and evolutionary biologists were eager to apply their knowledge and skills to grain agriculture. Billions of dollars' worth of research results, accumulated from the previous one hundred fifty years, could now be *applied* to the restoration of agriculture. The scientists had the software for the new hardware. The long path to a sustainable grain agriculture, modeled on wild systems, became clearer.

Imaginations went wild. Rangeland ecologists used grazing livestock along with fire as management tools. Feedlots for cattle, chickens, hogs, and turkeys emptied as animals returned to the farm to eat and deposit their urine and manure along with sheep and goats. Even camels were used in places for brush control. People with naturalist bents, but who had abhorred farming, became farmers. They loved their farming and were able to witness wildlife returned to rural areas.

In 2018, a legume-Kernza biculture was stabilized by biological nitrogen fixation, the same year that key nutrient-supplying and pathogen-suppressing roles of the soil microbiome were identified in perennial polycultures. That caught the attention of the scientific community. Soil carbon sequestration rates were modeled by 2020.

Two years later, the soil microbiome was intentionally managed for optimal health through crop breeding, inoculation, and fire and grazing. In 2025, a stable tri-culture was sustained by biological nitrogen fixation, leading to overyielding, which then became the rule.

Soon after, perennial sorghum became well established in sub-Saharan Africa, sustained by biological nitrogen fixation.

Perennial wheat hybrids expanded due to dozens of trials throughout Asia, Africa, Latin America, Europe, and the United States. Because of the work of plant pathologists and ecologists, by 2030 crop diversity was used to effectively regulate pests on field and landscape scales. Oil seed crops such as sunflower hybrids and *Silphium* were expanding their range. Millions of acres provided visible evidence for the new paradigm: a soil alive with wildness.

## The Land Needed Farmers

An agricultural historian at the University of Kansas writing in the mid-twenty-first century began a concluding paragraph with the following: "There is still much work to do." Her book detailed the institutionalization of the paradigm, describing how it happened, the courses that were taught, the research disappointments and breakthroughs, adoption by farmers, and so on. She described *ecological intensification* as an "information intensive" paradigm, one that had replaced the *energy intensive* industrial approach. By information, she meant a combination of human knowledge and the DNA of individual plants and their interaction with the rest of the biota. She described the various releases of the new perennials, field trials of species ensembles, and how those species made their way to the land.

The book clearly shows that a long-expected and major challenge had been met. But the new paradigm demanded more farmers. Where would they come from? Land ownership had become increasingly concentrated in the expanding fossil fuel era. It was deemed necessary to keep all current farmers in farming, even the bad ones. Nearly all farmers embraced the new paradigm as a compelling alternative. Their existing cultural knowledge was invaluable, and their know-how was being learned by new farmers who had grown up in cities and suburbs—farmers with countless limitations, but ready to try. They stuck with it because they found farm life satisfying, which was a good and timely thing, because the industrial era was about to close.

The Berry Center and networks of liberal arts colleges complemented the perennial polycultural research conducted at research universities, where programs were developed around the new agriculture. More schools joined by the year. Kansas Wesleyan, near The Land Institute, soon added an Ecospheric Studies program—the first of many.

What helped the young farmers "hang in there," as one put it, was that most of them had a broad liberal arts education that complimented the multidisciplinary thinking embedded in the new paradigm. Universities began to add entire colleges to accommodate schools of engineering, arts and sciences, and education. The University of Kansas and Kansas State University created *Schools of Ecospheric Studies*, filling catalogs with innovative courses. There were courses for would-be farmers and for majors in public policy, agriculture history, sociology, psychology, and the history of science. Humankind's role in changing the face of the earth was the organizing question and the moral centerpiece for these schools.

As more and more people began to see the world in more of its wholeness, the word *environment* fell from use, as did the term *biosphere*. As students learned to see through an ecospheric lens, they developed a coherent philosophical view that "biology by itself is incomplete," that unified ecological systems confer the properties we call life. The environment was no longer regarded as "out there," separate, part of *our* natural heritage, something we own, or a *thing* in the old subject-object dualism. Similar to the way in which Copernicus upended the commonsense notion of his time that the sun moved around the earth, we, who once saw the environment as something we acted upon, now see all organisms, including ourselves, as enclosed within a "miraculous skin."

Unfortunately, we in the West, for too long, too deeply, were descendants of those European ancestors who, around 1600, gave us the Enlightenment, gave us a way to know. For more than four hundred years, we have operated under the belief that a major way of knowing is to break a problem apart, to be reductive, to place priority on the part over the whole.

What we are advancing now is not exactly new. Greeks and Romans alike believed in universal orders of organization, and those orders were regarded as more important than individual organisms. The Greek theory of natural science held from Plato to the Stoics, whose worldview carried over to Romans such as Cicero (Rowe 1990, 49). Leonardo da Vinci was a late holdout of this outlook, but it was mostly plowed under during the Middle Ages. A few remnants survived to grow seeds among the nineteenth-century Romantics in Europe and North America, who in turn provided the philosophical framework for a growing number of modern conservationists. And now, finally, that is our worldview.

## *In Retrospect*

People in thousands of formal discussions, in classes and in seminars all over our ecosphere, are now pondering the human/nature split. Most scholars agree that it likely began with grain agriculture, which made civilization possible. Grain calories became more plentiful than calories from gathering and hunting, and grains can be easily stored. But here is the deeper source: Once humans adopted annual grain agriculture, they had to adopt the idea that nature is to be subdued. A favorable seedbed needed to be established if the seeds were to germinate. Following germination and growth, weeds had to be removed to ensure that the crop would thrive to harvest. Looked at this way, our human/nature dualism, our primary intellectual and philosophical burden, happened in innocence with our need to eat. But now we realize—or perhaps, better said, realize afresh—that nature is a necessary counselor, the standard and measure by which we are compelled to recognize our limits and act appropriately. If we can combine our exercise of limits with our adoption of nature as standard and measure, beginning with agriculture, we have a good shot at a new beginning . . . for everything.

There were those who presaged this time of new beginnings. I can do no better than to quote Professor Don Worster, who wrote the following:

> To conserve that evolutionary heritage is to focus our attention on the long history of the struggle of life on this planet. In recent centuries we have had our eyes fixed almost exclusively on the future and the potential affluence it can offer our aspiring species. Now it is time to learn to look backward more of the time and, from an appreciation of that past, learn humility in the presence of an achievement that overshadows all our technology, all our wealth, all our ingenuity, and all our human aspirations.
>
> To conserve that heritage is to put other values than economic ones first

in our priorities: the value of natural beauty, the value of respectfulness in the presence of what we have not created, and above all the value of life, itself, a phenomenon that even now, with all our intelligence, we cannot really explain. (155)

Those of us on Earth Island have a common mission: to protect it as an ecosphere. The effort is threefold: protect Earth's crust, protect Earth's waters, protect Earth's atmosphere. We rank Ecosphere protection as the highest of all callings for the following reasons:

1. It holds precedence in *time*. It was here long before we were.
2. It holds precedence in *inclusiveness*. We are embedded within it.
3. It holds precedence in *complexity* of organization. It is far more complex than we are.
4. It is more *creative*. Its evolutionary creativity has given rise to all biota, including us.
5. It has precedence in *diversity*.

In our educational efforts on Earth Island, we teach the young that nothing is more important than to comprehend the overarching supraorganismic reality we call the Ecosphere. To act on that ecology—to work with and cultivate this wild process—is central.

## References

Bacon, Francis. *New Atlantis*. In *The Works of Francis Bacon—Volume 3: Philosophical Works 3*, edited by James Spedding, Robert Leslie Ellis, and Douglas Denon Heath. First published 1857. Reissue; New York: Cambridge University Press, 2011.

Berry, Wendell. "The Culture of Agriculture." Presentation given at the Agriculture for a Small Planet Symposium, July 1, 1974, Gonzaga University, Spokane, WA. http://www.centerforneweconomics.org/e-newsletters/wendell-berry-culture-and-agriculture.

Rowe, Stan. *Home Place: Essays on Ecology*. Toronto: NeWest, 1990.

Worster, Donald. *The Wealth of Nature: Environmental History and the Ecological Imagination*. New York: Oxford University Press, 1993.

\* EPILOGUE \*

# WILD PARTNERSHIP

## *A Conversation with Roderick Frazier Nash*

### John Hausdoerffer

At first glance you may think Island Civilization is crazy and impossible, but don't stop with criticism. . . . If you disagree with some or all of my vision of an Island Civilization, put forward your own ideas about how our species should occupy this planet in a thousand years. —Roderick Frazier Nash (2015)[1]

FEBRUARY 22

February in Colorado's Elk Mountains can bring the sunshine that inspired Crested Butte's infamous naked skiers, or it can drop ten feet of snow in a week. My seven-year-old daughter Atalaya and I ride Crested Butte's High Lift to "The Headwall" at twelve thousand feet. We look toward the Maroon Bells Wilderness Area that divides our quiet valley from the buzz of Aspen sixteen miles (yet a six-hour mountain drive) to the north. Shafts of sunlight sneak through the beginning flurries of a predicted blizzard. The sky is wild, with a will of its own indifferent to mine. This seems fitting—at the bottom of the mountain, I am to meet the wilderness historian and philosopher Roderick Frazier Nash.

When not navigating his tugboat to Alaska, running a river in Utah, or hiking above Santa Barbara, Rod lives in Crested Butte, still skiing the steepest chutes at age seventy-six. In 1964, the same year the Wilderness Act was passed into law, Rod defended a dissertation that became his 1967 classic *Wilderness and the American Mind*. In *Wilderness*, Nash makes a bold claim: the idea of wilderness—once a way of naming the sinful domain of uncultivated land and beasts and now a sacred site of human and ecological regeneration—is an American revolution. This revolution, for Nash, rivals American independence from Britain and the emancipation of slaves during the Civil War, both in its fundamental shift in values and in its subsequent global impact. Nash's *Wilderness and the American Mind* gifted the wilderness revolution with a deep and mythic history. The book itself has become one of the forces of the wilderness movement.

Before meeting with Rod, I feel compelled to take in the 360-degree view from Headwall, scanning the Fossil Ridge, Raggeds, West Elk, Maroon Bells, Powderhorn, and Uncompahgre Wilderness Areas, all visible from our two-mile-high perch. Atalaya, looking down upon these wilderness areas that she hikes with me every summer, asks me what wilderness is. Having spent much of my career studying the wilderness idea, and having spent the previous week reviewing essays for this collection, I still do not have a good answer.

Knowing that kids understand more than we realize, I wonder where to begin. Do I detail Robert Michael Pyle's entire continuum, from his Denver ditch to the far reaches of "The Unknown"? Do I explain the differences between *Wilderness* (land zoned to protect habitat, to reduce human industrialized and mechanized presence, and to offer inspiration for those humans who can and want to seek it in such places) versus *wildness* (any self-organizing, self-renewing being, community, or system)? Or do I tell her what Enrique Salmón might say, that there are cultures without a word for *wild* because they do not differentiate "beyond and between the human and non-human worlds," or Gimiwan's view that wilderness is simply "home," or Jeff Grignon's and Robin Kimmerer's reminder that what we call "Wilderness" is often the human-shaped basis of livelihood within a people's homeland—"a carefully nurtured web of reciprocity between people and land"? Do I go further to tell Atalaya of concerns that wilderness can erase from memory those indigenous homelands? Do I tell her that some critics have said the wilderness idea itself has been a tool of colonialism, or that thinkers like Laura Watt worry our pristine wilderness ideologies still justify "aggressively pushing out long-established local uses?" Do I tell her what John Tallmadge might say, that the Maroon Bells Wilderness we gaze upon is not the only "complex adaptive system" to love and that places like Cincinnati are equally places where we can encounter nature in its "vivid, wild personhood"? Do I tell her about Gavin Van Horn's Greencorps Chicago friends' hope for the urban wild as a place of "peace, serenity, relief, fascination, and clarity of mind" in the midst of urban poverty and violence; or do I tell her about Michael Bryson's and Michael Howard's view of Chicago's South Side and Minnesota's Boundary Waters as reciprocally necessary to healthy communities everywhere? Or, still thinking of that same counterintuitive yet intricate Chicago wildness, do I share ecologist Seth Magle's question: "If the complexity is there . . . why don't we already think of these places as wild?" Do I tell my daughter what Brooke Hecht might say, that I love all places in which "the landscape itself is a story of life taking a chance—on the edge"? Do I tell her that the most sacred family stories, like those of Aaron Abeyta's grandfather, emerge from working within and carving a life out of wild land? Do I tell her what Rob Dunn might say, that the wild and "belligerent gods of rebirth" come to life as much in Atalaya's backyard compost pile as in the subalpine forest sweeping below us, and that, as Margo Farnsworth reminds us, "waste equals food" for everything from decomposers to urban-industrial visionaries? Do I explain to her the concept of the Anthropocene—the notion that humans now drive and are entangled in all environmental change—and thus pose to her Julianne Warren's question: "What is this making of wilderness, of Earth integrated and whole?" Or do I tell her what Roderick Frazier Nash, awaiting my arrival at the base of the mountain, might say—that our species threatens to "take down more than half the planet's wild species in a century" and that "with the wild goes [both the habitat for and] the existence rights of most nonhuman life on earth" (2014, 380–81)? All are true views of the wild, and I have yet to fully process what *wildness* means when one combines all these perspectives, let alone when one tries to translate such a collective view for a seven-year-old.

Respecting Atalaya's emerging understanding of a complex world, I almost tell her that wilderness is more about human potential than about pristine land. Smelling the dampening air until the cold freezes my nostrils, I almost explain that these wilderness areas depict the story of people deciding to slow themselves down before taking everything, to engage the world with humility rather than just desire. Instead, I simply tell her that these areas that surround her home come from our hope to share the land with all the other species with which we evolved. Atalaya blinks at me through her goggles and simply says, "I am glad animals have places like that to live." Rod would like her.

Before I can add another layer to Atalaya's comment, she asks if we can ski down to her four-year-old sister, Sol, waiting with a hot-chocolate mustache to share stories of a morning looking for snow fairies while gliding through the white world of "Bambino," her favorite forested trail. Atalaya and I tip our skis down the steeps of Headwall.

I am nervous as I walk to the bar. My personal understanding of wilderness owes a great debt to Rod Nash's books. In the 1990s, I always carried a torn copy of *Wilderness and the American Mind* while I ran a wilderness trail crew during the week and backpacked into every wilderness area I could access on weekends. Mostly, I am nervous because I want to talk with him about this collection, *Wildness,* and to learn from his reaction to a book project that qualifies the word *wilderness* with a wide continuum of notions of *wildness*. After all, in my writing and speaking, I have shared the concerns of historian William Cronon. Cronon warns us that the pristine *wilderness* idea distracts the environmental movement from fighting for *wildness*—"the autonomy and otherness of the things and creatures"— where we live and especially from struggling for environmental justice in exploited communities (89). Yet I bring to Rod my deliberate intention to find what is salvageable about the long-deconstructed wilderness idea, even as I continue to take to heart the unjust impacts of "wilderness" around the world—from the removal of the Bannock people to create Yellowstone Park in 1872, to the displacement of the Maasi in the late twentieth century, to the recent removal of sustainable food operations to develop a wilderness area in Point Reyes National Seashore just one month before this winter day in Crested Butte.

Rod has an inviting smile, easing my concerns. We sit and look outside, smiling as the blizzard finally arrives. We begin with a concern we share— that we have grown weary of "the debate" (reemerging during the fiftieth anniversary of the Wilderness Act in 2014) over whether wilderness is a distracting social construct or a vital natural reality. Rod reminds me that the word *wild* comes from the word *will* and that something wild is something that has will. We share a hope that diverse tributaries of the environmental movement (if such a movement exists) can find a common ground centered on wilderness's root meaning as "self-willed land." I share my view that "self-will" (as opposed to *selfish*-will) is as important a concept to wilderness movements as it is for working landscape movements (seeking to sustain the independent "self-will" of landowners who build resilience on their property) and for environmental justice movements (seeking to dignify and fight for the "self-will" of peoples excluded from choices about resource extraction and waste in their communities). I conclude, claiming, "Rod, nonselfish *self-will* is both the basis of wilderness and the best hope for the possibility of *wildness* becoming a unifying rather than a divisive worldview across the environmental spectrum."

Rod smiles at my optimism, reminding me that he has connected wilderness with social justice for decades. He speaks of his foundational 1989 book *The Rights of Nature*, in which he sees the "green world" as being "oppressed by the same exploitative, hierarchical values and institutions that once denied rights to slaves and continue to oppress many women, racial minorities, and laborers of all colors" (212). "John," he explains, "when you break a horse, what is broken? His will. Haven't we as a species substituted our will for that of many wild creatures, even when our cultivation is well intended?" Rod pauses while I try to imagine the moment a horse is "broken" and then shares his admiration for a 1963 essay by Wilderness Act architect Howard Zahniser: "Zahniser asked us to decide whether we will become gardeners or guardians on this earth. Gardeners still need to control; gardeners still need to substitute their will for that of many wild creatures. I decided then that I was a guardian of the will of other species."

Moved by the moral clarity of the "guardian" concept, I also worry that, already, Rod is creating a dichotomy between wilderness and civilization, guardians and gardeners. "But aren't there ways in which our species has blurred those boundaries?" I ask. "Aren't we capable of cocreative *partnerships* with the natural world, in which through the production of our livelihood we and the land together can create *more* biodiversity, build more resilience, and enhance the land's 'capacity for self-renewal'? Can't we learn from human communities who have coproduced wildness? Why can't we learn from those societies to become—again—something like a keystone species, cocreating with nature a necessary human niche rather than simply keeping ourselves out?"

Rod replies, "Not with seven billion people we can't, not within our industrial economy. But there is a bigger concern here in what you are saying. If we stick with my slavery analogy, your 'cocreative' vision between humans and the land is the equivalent of paternalism, in which the slaveowner argued that the slave could not have his own will and would still need to 'cocreate' his life under the guidance of his master rather than to demand his freedom. Your cocreative vision is simply a nicer way of keeping the land's will from that freedom."

## FEBRUARY 23

It had snowed ten inches that night, so I head to the mountain early the next morning in hopes of skiing untracked powder. Sitting on the ski bus, still unsettled by Rod's slavery analogy, I send him a message from my phone: "I understand your concern that a human 'cocreator' role with nature, given our history of domination, will only lead to a more palatable form of land slavery, not unlike how paternalism dangerously considered slavery a 'positive good' in the mid-nineteenth century." Thinking of Courtney White's and Joel Salatin's evidence of livestock building ecological resilience, I continue writing. "I see 'cocreator' as rooted in a mutual and equal partnership between human and more-than-human self-will, augmenting rather than violating the land's freedom. And reducing any mutual partnership with the land's agency down to something as violent as paternalism troubles me. We agree 90 percent on the value of wilderness, but these nuances are important to keep talking about!"

By the time the bus reemerges from the Internet data wilderness of Taylor Canyon, opening views of fresh snow on Paradise Divide, Rod has sent an e-mail reply: "I'm not as confident as you that the 'mutual partnership' between nature and technological (modern) human civilization will prove to be very helpful to the land's self-will, as it is still a form of control. That's why I suggest a cohabitation of the planet." Rod's term is in contrast to a term I have used with him—coinhabitation. Co*inhabitation* is the idea that through inhabiting the land with care, like Curt Meine's Driftless region having "grown wilder" thanks in part to farmers like Joseph Haugen, we can build ecological resilience. Co*habitation*, on the other hand, is Rod's idea that humans can only live ethically if wilderness is completely and forever separate. I agree with Rod on the importance of wilderness zoning on public lands, especially to curb mass extinction and to limit our fever for turning life into commodities, but I am not sure I agree with his view of eternal separation between humans and wilderness. Rod's words collide with those of the poet David Rothman, with whom I skied in this same basin a month earlier and who challenges his readers to "love a greater wildness of which we are always already part."

I finish the bus ride and ski one of my favorite stashes of fresh powder on a trail called Morning Glory. The snow is deep yet light in the steep trees behind the spot where Atalaya and I stood yesterday. Knowing this is one of Rod's favorite trails, I look for ski tracks to see if he has been here today. All signs of skiers from weeks past are erased by blankets of fresh snow. Powder days tap into a deep cultural mythos, not unlike the unreal lure of "pristine" land—skiing the first turns after a storm summons primordial feelings of being the first human on earth, and I experience this anew. Dropping into snow up to my waist, with each turn feeling as if I am walking down clouds, I ski to the lodge.

My thawing fingers tap my thoughts into the phone, replying to Rod: "I like your term 'cohabitation' a lot, Rod, as a tactic for protecting current wilderness areas. But what does that look like as a global strategy for the future wildness of the planet? One postindustrial example that fits my terms of co*in*habitation *and* cocreation through shared self-will is Wes Jackson's work at the Land Institute in Kansas. As you likely know, he is developing a 'perennial polyculture' that restores the prairie's self-organizing ecological function while providing food and fuel on a major scale. He believes this new food system, by end of century, will require and cause a revolution of ecological consciousness beyond its accomplishments of reducing soil erosion, pesticides, carbon emissions, etc. The potential is there for a *cocreative partnership that could result in more self-will for the land* rather than a new version of paternalistic 'control'—the prairie's long-collapsed self-organizing (wild) function could be reenlivened by our intelligent production of our food and fuel."

When I check my phone after a few more runs, I find this reply from Rod: "I like a lot of what Mr. Jackson is doing. As you know, my essay 'Island Civilization' looks a thousand years out to a way of inhabiting Earth that does not depend on agriculture/grazing as we know it now. By then we could fabricate our complete needs in the islands and really share the (now largely self-willed in 3015) planet with other species. I'm looking to a future in which 'nature' does not provide our food/fuel—we do it with a high but benign technology. Until then, I'm thinking even Mr. Jackson's good idea deprives nature of its will and involves a degree of control and (to put it perhaps too bluntly) constitutes a form of slavery. He is a gardener and not a guardian. As you know I favor humans backing off, standing down, using their tech smarts to build their habitats and leave the rest of the now-wild (willed) planet to its own devices and evolutionary futures. This vision (Martin Luther King would say 'dream') does not have much room for your relative wild. With the Islands, I'm moving beyond your cocreation, humans-as-part-of-nature scenario."

I do not reply until 10:00 that night. Despite tired ski legs burning as hot as the fire by my reading chair, I decide to reread Nash's "Island Civilization" before replying. Since "Island's" first appearance in a 1991 issue of *Wild Earth,* Nash has hoped to present something he is *for* after decades of looking back at what the environmental movement was *against*. Lamenting that wilderness has become "an endangered geographical species," "Island Civilization" looks ahead a thousand years and imagines that human ingenuity will be advanced enough by the fourth millennium to support a newly reduced human population occupying "several hundred concentrated habitats perhaps a hundred miles in diameter" (382). I reflect on how overpopulation arguments in the past have suggested suppressing the rights of our neighbors in the developing world. While recognizing that "Island Civilization brings some loss of freedom," Rod insists this reduction can come from moral choices in response to the tough lessons of ecological limitations rather than a dramatic oppression of free will (383). Human technology, in Nash's vision of the future, will allow for all food, energy, waste, and cultural activities to take place in those islands. This will allow for the rest of the planet, that which is outside these islands, to thrive as wilderness.

Enlivened by the essay, despite my physical fatigue, I type my late-night reply: "It took us just over twenty-four hours, but we found it, Rod, the spot where we diverge. Maybe 'self-will' as the definition of *wildness* discourages the partnerships that are possible. If we are going to use 'self-will,' don't we need to question the problematic Western roots of the individualized 'self' and dominating 'will' before applying it to how forests and rivers function? Partnerships must lie as a third way beyond both slavery and absolute freedom, and wilderness pushes me to be a partner (with forbearance). But it also pushes me to find how my choices enhance *shared self-will* between humans and land, beyond the *'visitor who does not remain'* of wilderness. That's not paternalism. That's 'give and take'—the oldest of ethical frameworks applied to the land. Relational Wild. Maybe that would be a better title for our *Wildness* collection."

FEBRUARY 24

Rod replies the following day:

> What you say here clicks on the light for me. I can see how you and Jackson fit into my dream. You are not forgetting your evolutionary roots even as your species moves ever more into a technological future. As an historian I am very aware that *my* father grew up in horse-and-buggy America.

So where could we be in a thousand years? I'd like to think not a wasteland or a total garden. We'll have the ability to shape our habitat and, hopefully, the ethics/restraint to leave the rest self-willed. And of course I use that term with reference to systems like rivers and mountains. You are right, that phrase *self-will needs explanation*. *Partnership* is worth exploring too, but I'm not optimistic about how successful our huge and greedy species could be for "partnering" with, say, salmon.

We have arrived at a crossroads, which I point out to Rod in the form of questions: "Must wildness—self-willed living systems as opposed to protected wilderness preserves—be, deliberately and unabashedly, *separate* from modern human livelihood in order to thrive? Have we proven that our activities, beyond those of small-scale indigenous communities, can never be designed to *enhance* and *coproduce* more self-will on the land, even across the most hopeful span of a thousand years?"

Rod seems to anticipate these questions, tapping his knowledge as a historian for his answers: "E. O. Wilson once said, 'Darwin's dice have rolled badly for earth.' What he is suggesting is that *Homo sapiens* really cannot be trusted with the earth. They screwed it up. There is very little in the historical record that suggests we can in the future be trusted to develop our ethics toward the land at the same pace with which we develop our technologies, and Island Civilization is about separation, which is painful because the garden scenario has long had an appeal in American nature writing. I understand, Island Civilization is a shocking idea, and that is what I am asking the readers to contemplate."

I deliberate carefully on Rod's lament for both lost wilderness and what he sees as the stubborn shortcomings of our evolutionary nature—we have proven that we cannot be trusted to partner without paternalism. But I pause, thinking of David Rothman's call for us to "love a greater wildness of which we are always already part," considering how Mistinguette Smith sees the "Black Wild" as "interstitial, the thing that's thriving in the in-between," and contemplating how Margot Higgins's beloved Wrangell–St. Elias National Park and Preserve opens an "inhabited wilderness" for the livelihoods of locals like Mark Vail. I am not sure I agree with E. O. Wilson's conclusion about Darwin's dice. Even further, the fact that Rod's Island Civilization must *continually choose* (even after the first choice to do so in a thousand years) a planetary design shaped around wilderness-promoting islands suggests that ethical change *can* move apace with technological hope.

## MARCH 17

One month later, on a March afternoon that feels more like June, Rod and I meet in Gunnison's Double Shot Bike Shop and Café. The place is busy. Open doors let in dry, warm spring air. Mountain bike trails are opening up in the desert regions of our valley to the west, and the clinking of mechanics preparing bikes dragged out of basement hibernation provides the hopeful sounds of spring—an ideal soundtrack for imagining our earth one thousand years from now. I have thought about Island Civilization a lot since his last e-mail.

Once again, I return to what Rod and I hold in common, the roots of self-will behind the term *wilderness*. I am reinspired by Rod's tenacious hope to liberate any system or species in possession of this "self-will." I empathize with his fears that even the good intentions of a "partnership ethic" will subtly draw wilderness back under human control, and it is hard to disagree that much of Western history has been dominated by a human will not to be trusted in any partnership with the self-will of nature. But I cannot help but think of Vandana Shiva's efforts in India to restore the dignity and autonomy of traditional nonchemical farming (on a large scale) and how this has been the basis for the "living energies" of humans and nature acting as the cocreative basis of the eighty thousand plants that humans have cultivated, shaping biodiverse living systems in the process. By thinking of Shiva, I recall Devon Peña's community acequia system in San Luis, Colorado. The democratic management of this ditch network for the sake of extensive food systems still leads to an increase of riparian habitat and connects people to food, forest, water, and alpine snowpack rising to fourteen thousand feet above the town.

But sitting in the coffee shop with Rod, I do not ask him to consider these directly. Instead of asking him to see our cocreative place outside the islands, within the wilderness, I ask him about his vision for the islands. "Rod, what is the place of wildness *within* your islands?" Rod looks intrigued, and I pause before explaining, "I am thinking of how people will need to value wildness within the islands on a daily level in order to value and continue to support wilderness outside the islands for millennia. For the good life to be enjoyed in these islands, if it is indeed a society that deeply values wilderness, there would need to be a strong presence of wildness within the islands, no?"

I ask this hoping that Rod will see the relevance of a more relative wildness within his islands, and then perhaps we could talk about whether a relative wildness could enhance the self-will of land outside the islands, today as well as in a thousand years. Rod seems to enjoy the playfulness of this conversation, as well as the opportunity to think more about life in his imagined islands. He reminds me that the "Island Civilization" essay pictures rivers running through islands and that each island would be its own unique bioregion to fit each person's nature aesthetic. I quiet my critical mind and enjoy this image—Roderick Frazier Nash's dream is very enticing.

A few days later, Rod sends a conclusive e-mail, having reflected on my question about the role of wildness within the islands:

> Thinking back again to that 1963 Howard Zahniser essay that called for us to be guardians over gardeners, maybe Zahniser did not create a wide enough spectrum of options. His world seems to have been very Manichean—*gardeners or guardians*. What about, Howard, some middle options? Fifty years later we need thinkers who will make room for more options from Howard's point of view and maybe my point of view. So I am glad the *Wildness* book is happening. But imagine if out there a thousand years from now, humans (now wonderfully empowered) have separated from the natural world and left it to be self-willed. When these humans of 3015 go outside their one-hundred-mile islands, they do so as "visitors who do not remain" and not as exploiters or controllers or gardeners—or, as we talked out, some might opt to live for years out there as neo-hunters and gatherers. The motive to respect the wild in this way comes from a new ethic that might or might not depend on invigoration from your "relational" contact. In the same way an antislavery ethic could exist strongly without any "relationship" with, say, black people (case in point, Thoreau). So, yeah, you and I had fun thinking about whales swimming freely alongside the habitable islands that could be partly in the ocean or salmon running up and down uncontrolled rivers that coursed through them. The views from the edges could be spectacular. And I would love to plan a months-long coast-to-coast trip with you in the restored wild matrix of 3015. But I don't see this as part of a "relative wild."

My conversations with Rod have focused me on the big question of this book, *Wildness*: Can we be trusted? Can we be needed? Even with Rod's growing openness to "more options" for the human place in the universe beyond gardeners or guardians, can we be trusted with a partnership with the more-than-human world? Or, at our best, must we simply decide we are naturally limited through preserving that which we cannot trust ourselves to call home—"sharing rather than partnering," as Rod says?

For me, it comes back to the answer I almost gave (and one day will give) my daughter Atalaya back in February, looking out from twelve thousand feet: "Wilderness areas tell the story of people deciding to slow themselves down before taking everything, to learn from the world with humility rather than just desire." But even this answer fails to satisfy my sense of the potential of wildness. I cannot shake this feeling: setting land aside does not completely capture the pinnacle of human choices possible, even for the sake of the wild. Rod invites us to "put forward" our own vision for how our species will occupy this planet, and my one-thousand-year, one-hundred-year, ten-year, and daily hope is that we can be wild partners.

## Note

1. Roderick Frazier Nash in discussion with John Hausdoerffer, February and March 2015.

## References

Nash, Roderick Frazier. *The Rights of Nature: A History of Environmental Ethics*. Madison: University of Wisconsin Press, 1989.

———. *Wilderness and the American Mind*. 5th edition. New Haven, CT: Yale University Press, 2014.

# *Permissions*

These credits are listed in the order in which the relevant contributions appear in the book.

Gary Snyder's "Wildfire News" first appeared in *This Present Moment: New Poems* (Counterpoint, 2015), 51. Copyright © 2015 by Gary Snyder. Reprinted by permission of Counterpoint.

Joel Salatin's "Order Versus Wildness" first appeared as "Order vs. Wildness" in *Acres U.S.A.* 43, no. 12 (December 2014). Reprinted with permission. www.acresusa.com.

Courtney White's "The Working Wilderness" first appeared as a chapter in *Revolution on the Range*, by Courtney White. Copyright © 2008 Courtney White. Reproduced by permission of Island Press, Washington, DC.

A variation of Rob Dunn's "The Whiskered God of Filth" first appeared as "What If God Were a Maggot?" on *The Scientific American—Guest Blog* (December 20, 2012).

A variation of Wes Jackson's "Earth Island: Prelude to a Eutopian History" first appeared as "The Great Awakening: Where We Are Headed with Agricultural Management" in *Catamaran Literary Reader* 9 (Spring 2015): 23–27.

# About the Contributors

AARON A. ABEYTA is a professor of English at Adams State University. He is the author of five books and the recipient of a Colorado Book Award and an American Book Award. He lives in Antonito, Colorado, his hometown, where he also serves as Mayor.

MICHAEL A. BRYSON is Professor of Humanities and Sustainability Studies at Roosevelt University and a research associate in science and education at the Field Museum of Natural History, both in Chicago, Illinois. He earned a BA in biology and English at Illinois Wesleyan University and a PhD in English from Stony Brook University. Bryson cofounded and directs Roosevelt's Sustainability Studies program and is a widely published author on sustainability education, urban nature writing, and the ecology of cities.

ROB DUNN is an ecologist and evolutionary biologist who studies and writes about the wild of our everyday lives, whether that means sacred groves in Ghanaian backyards or pillowcases in midwestern houses. He has written three books: *Every Living Thing*, *The Wild Life of Our Bodies*, and most recently, *The Man Who Touched His Own Heart*.

MARGO FARNSWORTH invites readers into nature, offering strategic ways to live with wild neighbors through biomimicry and other practical methods. Her work has appeared in *Outdoor Living*, *TreeHugger*, and *Earthlines*, as well as the *City Creatures Blog*. She is currently writing a book on how corporations discover and use biomimicry. She works from her Missouri farm, which she shares with her husband, assorted pets, and wildlife.

JEFF GRIGNON is a father, grandfather, husband, and enrolled member of the Menominee Tribe of Wisconsin. A lifelong student of the environment, having eighteen years of western and southern fire experience, he has now accepted the honor and responsibility of aiding in the regeneration or the giving back of what the Menominee forest has offered. His responsibilities as a tribal member also include the preservation and protection of cultural habitational areas and the stories they continue to tell.

JOHN HAUSDOERFFER is executive director of the Center for Environment and Sustainability at Western State Colorado University, where he is also a professor of philosophy and environment and sustainability. Hausdoerffer is author of the 2009 book *Catlin's Lament: Indians, Manifest Destiny, and the Ethics of Nature* as well as editor of Aaron Abeyta's 2014 collection *Letters from the Headwaters*. When seeking inspiration and (relative) wildness, Hausdoerffer skis with his wife and two daughters from a reclaimed-wood shack they built above Pitkin, Colorado.

BROOKE HECHT is a human with an avid interest in *what it means to be human*. She explores her humanity as a mother, wife, and president of the Center for Humans and Nature. She received her PhD from Yale University's School of Forestry and Environmental Studies, where her research focused on ecosystem edges and the stories they tell about ecosystem history, resilience, and human/nature relationships. Hecht continues to explore the edges — of how we understand our moral obligations to each other and the whole community of life.

MARGOT HIGGINS is a visiting professor in the Environmental Studies Department at Macalester. She has had the great privilege of spending the last decade living and conducting research in the socioecological setting of Alaska's Wrangell-St. Elias National Park and Preserve. In addition to her history of scholarship and activism within the academy, she has worked professionally as a journalist, documentary filmmaker, campaign organizer, and nonprofit leader.

# ABOUT THE CONTRIBUTORS

MICHAEL HOWARD's life passion is to improve the quality of life for citizens of Chicago's Fuller Park community, both financially and environmentally. As founder and CEO of Fuller Park Community Development (FPCD), he works to address housing, education, and environmental issues that have kept this generally African American and low-income community in poverty and disrepair.

WES JACKSON is cofounder and president of the Land Institute in Salina, Kansas. He is the author of several books and papers, including *Consulting the Genius of the Place: An Ecological Approach to a New Agriculture, New Roots for Agriculture,* and *Becoming Native to This Place.* Jackson is the recipient of several awards, including the Pew Conservation Scholars award, a MacArthur Fellowship, and the Right Livelihood Award (Stockholm), known as the "Alternative Nobel Prize."

ROBIN WALL KIMMERER is a mother, scientist, writer, and Distinguished Teaching Professor of Environmental Biology at the SUNY College of Environmental Science and Forestry in Syracuse, New York. Kimmerer is an enrolled member of the Citizen Band Potawatomi. Her writings include numerous scientific articles and the books *Gathering Moss,* which was awarded the John Burroughs Medal for nature writing in 2005, and *Braiding Sweetgrass: Indigenous Wisdom, Scientific Knowledge and the Teachings of Plants.*

SETH MAGLE has studied animals in urban areas for almost twenty years, including research on prairie dogs living in median strips, white-tailed deer in urbanizing landscapes, coyotes in America's third-largest city, and Canada lynx trying to dodge traffic on eight-lane highways. He strongly believes that if rare and imperiled species are to be conserved in our modern world, we must understand and mitigate all potential impacts of cities on wildlife, and to that end, he engages in studies that span a broad range of scientific disciplines, including behavioral ecology, conservation genetics, landscape ecology, environmental education, and human dimensions of wildlife. His vision is to help create a world in which urban ecosystems represent an important component of the worldwide conservation of biodiversity.

## ABOUT THE CONTRIBUTORS

CURT MEINE is a conservation biologist, historian, and writer who serves as Senior Fellow with the Aldo Leopold Foundation and the Center for Humans and Nature and as associate adjunct professor at the University of Wisconsin–Madison. He has written and edited several books, including *Aldo Leopold: His Life and Work* and *Correction Lines: Essays on Land, Leopold, and Conservation*, and is the on-screen guide in the documentary film *Green Fire: Aldo Leopold and a Land Ethic for Our Time*.

DEVON G. PEÑA is a professor of American ethnic studies, anthropology, and environmental studies at the University of Washington. He is also the founder and president of the Acequia Institute, a charitable foundation that supports the environmental and food justice movements and operates a 181-acre acequia farm in Colorado. The author of numerous peer-reviewed articles, several award-winning books, three edited volumes, and senior editor of two Oxford encyclopedias, Dr. Peña is coeditor of the forthcoming book *Mexican-Origin Foods, Foodways, and Social Movements: A Decolonial Reader*, to be published by the University of Arkansas Press. He is also completing work on his next monograph, *The Last Common: Endangered Lands and Disappeared People in the Politics of Place*, a thirty-year "deep" ethnography of land and water rights struggles in southern Colorado.

ROBERT MICHAEL PYLE is the author of twenty books of prose and poetry, including a flight of butterfly books. His awards include the John Burroughs Medal (*Wintergreen*), a Guggenheim Fellowship (*Where Bigfoot Walks*), and the National Outdoor Book Award (*Sky Time in Gray's River*). Pyle has taught as Kittredge Distinguished Visiting Writer at the University of Montana and in place-based workshops from Tasmania to Tajikistan. He writes and studies natural history from an old Swedish homestead in southwesternmost Washington.

DAVID J. ROTHMAN's most recent books include two volumes of poetry, *The Book of Catapults* (White Violet Press) and *Part of the Darkness* (Entasis Press), and a collection of creative nonfiction, *Living the Life: Tales from America's Mountains and Ski Towns* (Conundrum Press), all published in 2013. He directs the

Graduate Program in Creative Writing at Western State Colorado University in Gunnison and lives nearby in Crested Butte. He represents the southwest region on the board of the Association of Writers and Writing Programs (AWP) and is the Resident Poet of Colorado Public Radio. In addition to continuing to write poems, he is completing a critical book on Robinson Jeffers, along with a scholarly work on literary prosody and a new collection of creative nonfiction about life in the mountains.

JOEL SALATIN is a third-generation beyond-organic farmer and author whose family owns and operates Polyface Farm in Virginia's Shenandoah Valley. The farm produces salad bar beef, pigaerator pork, pastured poultry, and forage-based rabbits and direct markets everything to five thousand families, fifty restaurants, and ten retail outlets. A prolific author, Salatin's nine books to date include both how-to and big-picture themes. The farm features prominently in Michael Pollan's *New York Times* bestseller *Omnivore's Dilemma* and the award-winning documentary *Food Inc.*

ENRIQUE SALMÓN is Rarámuri. He is the author of *Eating the Landscape: American Indian Stories of Food, Identity, and Resilience.* He is professor of American Indian studies at California State University East Bay in Hayward, California, where he teaches American Indian and ethnic studies. Salmón researches, contemplates and writes about American Indian relationships to place and how some indigenous communities are working toward mitigating the impacts of climate changes based on ancestral ecological knowledge.

VANDANA SHIVA trained as a physicist at the University of Punjab and completed her PhD at the University of Western Ontario, Canada. She later shifted to interdisciplinary research in science, technology, and environmental policy. Dr. Shiva has contributed in fundamental ways to changing the practice and paradigms of agriculture and food. Her books *The Violence of the Green Revolution* and *Monocultures of the Mind* pose essential challenges to the dominant paradigm of nonsustainable, industrial agriculture. Through her books *Biopiracy*, *Stolen Harvest*, and *Water Wars*, Dr. Shiva has made visible the social, economic, and ecological costs of corporate-led globalization. In November 2010, *Forbes* identified Dr. Shiva as one of the "Seven Most Powerful Women on the Globe." Dr. Shiva advises governments worldwide and is currently working with the Government of Bhutan to make Bhutan 100 percent organic.

MISTINGUETTE SMITH loves stories, data, unlikely intersections, and social activism. She is the founding director of the Black/Land Project, which synthesizes contemporary oral narratives, participatory action research, and tools for community organizing. Formally trained as both a poet and a bureaucrat, she teaches visionary, skillful leadership in many places, including the University of Vermont Rubenstein School of the Environment. She lives 1,500 feet from the Connecticut River.

GARY SNYDER was born in San Francisco in 1930 and is a writer, Buddhist, forest landowner, and bioregionalist. He has published eighteen books of poetry and prose, including the poems of *Turtle Island* (Pulitzer Prize, 1975), *Mountains and Rivers without End* (Bollingen Prize, 1997), and the paradigm-shifting essays of *The Practice of the Wild* (1990). He has been a logger, a forest firefighter, a lookout, a professor, and a Zen student in Japan. He lives in the Yuba watershed of the Northern Sierra.

Fleeing a New Jersey childhood, JOHN TALLMADGE received an indoor education from Dartmouth and Yale and an outdoor education from the White Mountains, the High Sierra, the Colorado Plateau, and the Boundary Waters. After teaching at the University of Utah and Carleton College, he settled in the Ohio Valley to raise children and explore urban nature while working with adult learners. He is the author of *Meeting the Tree of Life: A Teacher's Path* and *The Cincinnati Arch: Learning from Nature in the City*, as well as numerous essays in natural history, ecocriticism, outdoor education, and environmental philosophy.

GAVIN VAN HORN is the director of Cultures of Conservation for the Center for Humans and Nature. He is the coeditor of *City Creatures: Animal Encounters in the Chicago Wilderness* (University of Chicago Press, 2015). He writes for, edits, and curates the *City Creatures Blog*. He is currently working on a book of creative nonfiction, *The Channel Coyotes*, which highlights various urban animals and the ways in which they can deepen human understandings of place.

JULIANNE LUTZ WARREN is author of *Aldo Leopold's Odyssey*, unfolding the ideas of this keenest of American ecologists about wildness, human culture, and Earth's health. Warren has a PhD in wildlife ecology. She won an NYU Martin Luther King Jr. Faculty Research Award for her work with students on climate justice. Currently, she serves as a fellow with the Center for Humans and Nature and writes in her home in Ithaca, New York, or a rented dry cabin in Fairbanks, Alaska.

LAURA ALICE WATT is a professor in the Department of Environmental Studies and Planning at Sonoma State University. She is an environmental historian interested in the ways in which protection and restoration of ecosystems affect the interactions of people and place over time. Her book *The Paradox of Preservation: Wilderness and Working Landscapes at Point Reyes National Seashore* was published by the University of California Press in 2016. She is also an avid photographer and sailor.

A former archaeologist and Sierra Club activist, COURTNEY WHITE dropped out of the "conflict industry" in 1997 to cofound the Quivira Coalition, a nonprofit dedicated to building bridges between ranchers, conservationists, and others around practices that improve economic and ecological resilience in western working landscapes. He is the author of *Revolution on the Range* (Island Press), *Grass, Soil, Hope* (Chelsea Green), *The Age of Consequences* (Counterpoint Press), and *2% Solutions for the Planet* (Chelsea Green). He lives in Santa Fe, New Mexico, with his family, two dogs, and four chickens.

# *Index*

Readers are encouraged to consult terms relating to *Wildness*: wild (in adjective and unspecified forms), wild (the wild), wilderness, Wilderness (proper noun form), Wilderness Act, wilderness areas, wildlife, and wildlife refuges, as well as these terms interspersed in titles. Similar concepts are sometimes found under two or more of these terms.

Abbey, Edward, 76, 178–80
Abeyta, Aaron, 123–34, 245, 261, 262
Abram, David, 178
Acequia Institute (San Luis, Colorado), 92, 264
acequia waterways, 19–20, 90–93, 96, 252
*ACRES USA* magazine, 45
Adams, Ansel, 60, 148
African Americans. *See* Eden Place Nature Center; Greencorps Chicago; slaves and slavery; "Wild Black Margins"
Agamben, Giorgio, 96, 98n6
agriculture: biodiversity threatened by, 234; butterfly endangerment as threat to, 19; and debt-bondage sharecropping, 139; economic development advanced by, 235; grain agriculture, 234–37, 240; Green Revolution in, 234–35; human dependency on, 249; mutualism in, 20; opposing factions in, 234–37;
oyster farming as a form of, 100–112; quality and quantity in, 236; and revitalization, 240–41; traditional vs. technological methods in, 47–48, 235–37; urban renewal vs., 139; and wildness, 47, 49, 235–36
air pollution, 171, 219
"*Akiing* Ethic, The: Seeking Ancestral Wildness beyond Aldo Leopold's Wilderness" (Hausdoerffer), 195–204
Alaska. *See* "Inhabiting the Alaskan Wild"
Alaska National Interest Lands Conservation Act (ANILCA), 115–16, 120–21
Alberta tar sand project, 18
Aldo Leopold Foundation, 264
"All Apologies" (Nirvana), 18
Allen, Craig, 84
*All Our Relations* (LaDuke), 196
Altar Valley (Arizona), 75, 81, 87
Alzheimer's disease, 231

*American Copper* (Ray), 16
American Indians, 27–29, 196, 265. *See also* Bannocks; First People; Indian Service (US); Maasis; Menominees; Native Americans; Navajos; Ojibwes; Tohono O'odham; Yakamas; Yukaghirs
Amish culture, 40–41
Anderson, Ray, 51–54, 58–59
ANILCA (Alaska National Interest Lands Conservation Act), 115–16, 120–21
Anishinaabe (American Indian nation), 7, 195, 198–203
"Answer, The" (Jeffers), 60–61
Anthropocene epoch, 7, 215–19, 222, 224, 245
Apaches (American/Mexican Indian Nation), 26–27
Assembly of All Beings, 150
Athabascans (American Indian nation), 116
*autopoiesis* (self-creation), 230

bacteria, 178, 193–94, 209, 231
Baer, Morley, 60
Bandelier National Monument and Wilderness (New Mexico), 84–85
Bannocks (American Indian nation), 246
Barker, Joel Arthur, 48
Basin (Massachusetts), 62–63
Baumeister, Dayna, 54–55
beekeeping, 47–48
Benedict, James, 13
Benyus, Janine, 54–55
Berry, Thomas, 178
Berry, Wendell, 139, 236
Berry Center, 239
Big Wild, 16
biodiversity: agriculture as threat to, 234; in cities, 156–57; and cultural diversity, 24; East Indian conceptions of, 230; in Hegewisch Marsh, 153–54; human communities in regions of, 25; human promotion of, 247; in managed lands, 172; in the Menominee Forest, 74; public responses to, 56; and the Rarámuri, 5, 24; in the Sierra Tarahumara, 24–25; in wildlife refuges, 172; women as drivers of, 231. *See also* diversity
biomass, 43–44, 46
biomes, 159, 231, 238
biomimicry, 261–62. *See also* permaculture
"Biomimicry: Business from the Wild" (Farnsworth), 50–59
*Biomimicry: Innovation Inspired by Nature* (Benyus), 54–55
Biomimicry Institute, 53
biophilia, 180
*Biophilia* (Wilson), 29
biopiracy, 97
biotic community, 196
Black/Land, 138–42, 266
black people, 137–43, 150, 253
black wild, 138, 142, 251
blended spaces, 30–31. *See also* cognitive blending
Boabeng-Fiema, Ghana, 189–90, 192
Booth, Michael, 208–10
Borgie Glen (Scotland), 23
"Born to be Wild" (Steppenwolf), 3
Boundary Waters (Minnesota), 169, 177, 184, 245, 266
bovine spongiform encephalopathy, 48–49
Bowe, Roger, 81
Bradford, John, 52–53, 56–59
Braungart, Michael, 54
Brower, David, 60
Brown, Stan and Ruth, 62–63
Brunetti, Jerry, 48

Bryson, Michael, 166–76, 245, 261
Buber, Martin, 180
Buenos Aires National Wildlife Refuge (New Mexico), 76, 85
"Building the Civilized Wild" (Magle), 156–65
Bureau of Land Management (US) (BLM), 87, 102
Burroughs, John, 183
Burton, John, 106
businesses reconciling with nature. See "Biomimicry: Business from the Wild"
butterflies, 14, 19, 43–44, 218, 230, 264

Cairngorms (mountains) (Scotland), 23
Caldwell Preserve (Cincinnati), 181
Callicott, Baird, 17
capitalism, 22, 89, 93, 96–97, 229
carbon emissions, 219, 249
Carter, Jimmy, 115
Carver, George Washington, 139
catfish as gods, 7, 192–94
Catlin, George, 196
Catton, Theodore, 115
Centers for Disease Control and Prevention (US), 182, 231
*Chabochi* (white people in Rarámuri language), 25–26
Chaco Culture National Historical Park (New Mexico), 81–83, 85
Chicago, Illinois: ecological restoration in, 145–46; forest preserves in and around, 145–53; South Side of, 6, 146, 148–50, 153, 166–76, 245; West Side of, 146–48; wild street conditions in, 148–49. See also "Building the Civilized Wild"; Eden Place Nature Center; Greencorps Chicago
Chicago Park District, 162
Chicago Wildlife Watch, 163
Chihuahua, Mexico, 24, 27

Christians, 1, 189–90, 192
Cicero (Roman philosopher), 240
Cimarron, New Mexico, 77–79
Cincinnati, Ohio, 6, 178–79, 181–82, 184, 245. *See also* Fernald (Ohio Superfund site)
cities. *See* "Building the Civilized Wild"; Chicago, Illinois; Cincinnati, Ohio; "Cultivating the Wild on Chicago's South Side"; Greencorps Chicago; "Healing the Urban Wild"; urban sprawl; "Wild Black Margins"
*Citizens United* (US Supreme Court ruling), 95
Civilian Conservation Corps, 34–35, 181
climate change and global warming: acequias threatened by, 98n2; in Alaska, 117, 121; awareness of, 233; cultural climate change, 69; and the Driftless area, 40; and the Menominee Forest, 73–74; and a post-fossil fuel economy, 229; record pace of, 216, 222
Clinton Forest Plan, 16
closed systems, 57
coal, 21, 125, 133, 229. *See also* fossil fuel development and consumption
cocreation, 7, 108, 202, 230–31, 247–49, 252
cognitive blending, 30. *See also* blended spaces
cohabitation and coinhabitation, 3, 91, 248–49
College of the Menominee Nation (CMN), 73
colobus monkeys as gods, 7, 189–90, 193
colonialism: colonial-capitalist binaries, 96–97; colonial dispossession, 93–94, 98n3; colonialization of knowledge, 201; colonial theft, 116; and individualism, 96;

colonialism (*continued*)
    and Leopold, 201; the meaning of wild/wilderness in, 5, 230, 244–45; Native people displaced in, 121; resisted, 93–95; settler colonialism, 98n3, 142; unpossessive presence vs., 92; and the wild, 138
Colorado. *See* acequia waterways; Crested Butte, Colorado; Culebra River/Peak/watershed; High Line Canal; Indian Peaks Wilderness; San Luis Valley; Willapa Hills
common market era, 19
community: acequia communities, 91–96; business creating, 59; conquering vs., 93, 196; Earth's community of life, 103; in Eden Place Nature Center, 168–74; and Greencorps Chicago, 153; and land, 19–22, 86, 153; Leopold on, 1–2; in the Menominee Forest, 70–74; necessity of, 3–4; in Point Reyes, 110, 117–20; self-renewal of, 1–2; Snyder on, 145; and wild processes, 3; of wild systems, 47
Congress (US), 16–17, 101–2, 104–5, 115
conservation: and business leadership, 52; of butterflies, 19; conservation biology, 97, 106; conservation movement, 25, 76–86; in Coon Valley Wisconsin, 33–35, 37, 40; ephemerality in, 109; ethics of, 77; and Greencorps Chicago, 149, 152; in India, 230; and Native cultures, 19; sports fishing vs., 20; and Ugandan sacred groves, 191; of water, 76; and Wrangell-St. Elias National Park, 118
conservationists, 2, 15, 17, 25, 40, 148, 156–58. *See also* Leopold, Aldo; "Working Wilderness, The"
Constitution (US), 95

Continuous Forest Inventory (CFI) sampling plots, 71
"Conundrum and Continuum: One Man's Wilderness, from a Ditch to the Dark Divide" (Pyle), 3, 12–23
Coon Valley (Wisconsin), 34–35, 37–38, 40–41
Corathers, Robin, 182
Coulthard, Glen, 93
Crested Butte, Colorado, 243–44, 246, 265
Cronon, William, 102, 108–9, 246
Culebra River/Peak/watershed (Colorado), 90–91, 94, 96, 98n1
"Cultivating the Wild" (Shiva), 228–32
"Cultivating the Wild on Chicago's South Side: Stories of People and Nature at Eden Place Nature Center" (Bryson and Howard), 166–76
Curry, Richard, 106, 111n4
cycles (natural and ecological), 51–59, 151, 202, 207

Dahl, Michael, 195–99, 202–3
dams, 22, 38, 77, 103, 105
Daniel, Sydney, 55
Dark Divide Wilderness Area (Washington), 5, 16
Darwin, Charles, 225, 225n1, 251
Davis-Stafford, Julia and Kim, 77, 79
death in nature, 54, 123–25, 126–29, 134
decomposers, 7, 193–94, 245
Dempsey, Jack, 130–33
Denver Water (utility), 19
*Desert Smells like Rain, The* (Nabhan), 19
Detroit, Michigan, 140–41
Dinosaur National Monument (Utah), 103
displacement, 116, 121, 138–39, 224, 246
diversity: acequias promoting, 92; and biomimicry, 55–56; cultural,

24; in the Ecosphere, 241; in ecotones, 142; in farmland, 44; genetic, 22; humans creating, 25; and land health, 81–83, 87; in the Menominee Forest, 74; as soil health indicator, 83; in urban areas, 162; wild systems as standard for, 7. *See also* biodiversity

Donne, John, 50

Douglas, William O., 15

Drakes Bay Oyster Company (DBOC), 100–112

Drakes Estero (estuary) (California), 100, 108

Driftless Area (Wisconsin), 36–40, 248. *See also* "Edge of Anomaly, The"

dumpage, 219–20

Dunn, Rob, 189–94, 245, 261

Dyson, Freeman, 183

Eaarth (McKibben's altered Earth), 215, 224

Earth First! (environmental activist group), 106

"Earth Island: Prelude to a Eutopian History" (Jackson), 233–41. *See also* "Island Civilization"

Earth's Operating Conditions, 53–54, 59

eco-farming, 5–6, 45–47, 49

*Ecological Conscience, The* (Leopold), 75

ecological health, 76, 79–80, 84, 87, 235–36

ecological illiteracy, 80

ecological intensification, 234–36, 238

ecological literacy, 176, 183

ecological resilience, 248, 267

ecological restoration, 145–47. *See also* "Building the Civilized Wild"; "Cultivating the Wild on Chicago's South Side"; "Healing the Urban Wild"

ecological systems. *See* ecosystems

ecology: economy vs., 96; of the Ecosphere, 241; ecosystem ecology, 205; human sovereignty vs., 97; kincentric ecology, 5; of rangeland health, 80; reconciliation ecology, 160; scienticized ecology, 97; traditional ecological knowledge, 70

economics: agriculture advancing, 235; alter-economies vs. capitalist economies, 96; of biomimicry, 50–59; in Coon Valley, Wisconsin, 35, 39; of disadvantaged communities, 145–55, 166–76; and the Drakes Bay Oyster Company, 100–112; ecology vs., 96; economic efficiency misinterpreted, 96; emotional relations to nature vs., 141; evolutionary heritage vs., 229; of fishing, 39; and the Ganges, 229; and land health, 87–88; and Leopold, 102; in Menominee culture, 69; of ownership, 141; of renewable energy, 229; transeconomic values, 197; and wildness, 6. *See also* colonialism

Ecosphere, 234, 237, 240–41

ecosystems: and biomimicry, 50–51; and colonialism, 97; culture/nature coupling in, 90; and the DBOC, 107–8; deterioration of, 84–86; of the Drakes Estero estuary, 108; of Earth Island, 239; ecotones of, 142; fire as beneficial in, 78; and Fuller Park, 168; and grain agriculture, 236–37; human interaction with, 20; and human psychology, 208; humans in, 20, 24–25, 28, 31, 70, 115; of Iceland, 205–13; natural vs. monocultural, 236–38; and ordered farming, 46; of rangelands, 80–81; self-willing bioregional ecosystems, 90, 93; Snyder on, 150; sustainability of, 28, 80

ecotones, 6, 138, 142–43. *See also* margins/marginality
Eden Place Nature Center (Fuller Park, Chicago), 166–76
"Edge of Anomaly, The" (Meine), 33–42
Eggers Woods (Chicago), 145–47
Egg Lake (Minnesota), 199–200
Eiseley, Loren, 60
*El Comal* (Colorado), 90, 98n1
Elder Tree Nations (in Menominee guardian legend), 67–69
elitism, 17
"elk calf, the" (Abeyta), 125–29
Ellis, Tyrone, 149, 151, 153
Elmore, Brenda, 149–50, 152
endangered species, 19, 80, 147, 158, 162–63, 250
*End of Nature, The* (McKibben), 215
energy, 229–32, 238, 252. *See also* coal; fossil fuel development and consumption; oil
Enlightenment (era), 233, 239
environment: of Alaska, 117; the built environment, 138, 140; definitions and meanings of, 31; of Detroit, 141; and the Ecosphere, 239; environmental ethics, 77; environmental exploitation, 84; environmental injustice, 168; environmental justice, 196, 235, 246–47; exploitation of, 84; holistic views of, 114; the human-environment interface, 25; human sovereignty vs., 97; Menominee respect for, 67, 74; subsistence from, 113–18
Environmental Action Committee (Marin County, California), 107
environmentalists, 75–84, 100–112, 124, 137–38, 168, 175
environmental movement, 25, 78–80, 86, 106, 156–57, 246, 249–50
"Epilogue: Wild Partnership: A Conservation with Roderick Frazier Nash" (Hausdoerffer), 243–54
Erikson, Erik, 225n1
ethics: and acequias, 96; "The *Akiing* Ethic," 195–204; of biomimicry, 56–57; challenges required by, 97; of conservation, 77; of cultural equality, 93–94; involving endangered species, 162–63; environmental ethics, 77; and knowledge, 77; "The Land Ethic," 77, 92–93, 196–97; of man vs. nature, 109, 253; partnership ethics, 252; profitability of, 79; among ranchers, 77; and time, 195; of woodland disturbance, 208
"Ethnobotany of the Menominee Indians, The" (Huron Smith), 71
eutopia. *See* "Earth Island"
extinction. *See* species extinction
Exxon Mobil, 217
Eysteinsson, Þröstur, 212–13, 213n1

farming: and acequias, 90–92; in America, 31; biodynamic farming, 139; black farmers, 174; and debt-bondage sharecropping, 139; ecofarming, 5–6, 45–47, 49; monoculture farming, 94, 216, 237; natural vs. chemical-based, 230–31, 252; as a necessity, 238–40; ordered farming, 44–47; returning to, 237–38; and slavery, 173–75; spiritual aspects of, 183, 198; Thoreau on, 183; traditional vs. technological methods in, 47–48, 230–31, 235–37, 252; in wildness, 49
Farnsworth, Margo, 50–59, 245, 261
Federal Wilderness Areas. *See* wilderness areas
feeding the world, 230–31, 233–34
Feinstein, Dianne, 109
fence-line contrasts, 81–82

Fernald (Ohio Superfund site), 182–84
fertilizers, 48, 113, 174, 230, 234–35, 237–38
fires and fire: beneficial effects of, 83; as controlled burn, 149; Great Yellowstone fire, 25; as a land management tool, 237; Ojibwe words for, 202; rest from, 81; routine fires, 45–46; "Wildfire News," 11; wildfires, 25, 149, 220
First People, 93–95, 222
fish and fishing, 20, 39, 91, 116, 162, 192–94. *See also* Drakes Bay Oyster Company
Fish and Wildlife Service (US), 102
Flint, Howard W., 17
Fluted Peak (Colorado), 89
foliar nitrogen, 207–12
*Food Inc.* (documentary film), 265
"Footprints in the Snow" (Benedict), 13
Foreman, Dave, 200–201
"Forest, The" (Leopold), 197
forest preserves, 145–53, 158, 161, 164, 169, 181
forests: Anishinaabeg knowledge of, 198; in Appalachia, 46; clear-cutting in, 15, 69, 94; Continuous Forest Inventory sampling plots, 71; contractions of, 190; disturbances of, 208–12; foliar nitrogen in, 207; of Ghana, 190–91; Gifford Pinchot National Forest, 22; Gila National Forest, 197–98; humans as beneficial to, 67–74; in Iceland, 205–9; Kaibab National Forest, 79–81; Leopold's conceptions of, 197; Menominee Forest, 6, 67–69, 71, 262; Mount Airy Forest, 181, 184; regeneration of, 6, 70–74, 262; as sacred groves, 7, 190–94; sustainability of, 72; the Unknown (Scottish forest area), 23, 244. *See also* logging; "On the Wild Edge in Iceland"; timber production

Forest Service (Iceland), 212
Forest Service (US), 22, 102–3
fossil fuel development and consumption: advent of, 28; awareness of, 233; and carbon emissions, 219, 249; and land ownership, 238; vs. living energy, 229; opposition to, 216–19, 224; products of, 57–58; wilderness areas affected by, 21. *See also* coal; gas (natural gas); greenhouse gases; oil
Franzen, Jonathan, 218–19
Friedman, Jerry, 106, 110n2
Friends of the Forest Preserves (Chicago), 149
Fuller Park (Chicago), 166–69, 175
Fuller Park Community Development (FPCD), 263
function-value controversy, 84–86

Gadzia, Kirk, 80–82
*Game Management* (Leopold), 87
Ganga (Hindu deity), 228
Ganges River, 228–29, 231–32
García, Reyes, 92
gardeners vs. guardians, 247, 249, 253
gas (natural gas), 21, 218, 229. *See also* fossil fuel development and consumption
Gaumukh glacier (India), 231–32
genetically modified organisms (GMOs), 48
genius of place, 5–6, 51
Ghana, 189–92
Gifford Pinchot National Forest (Washington), 22
Gila National Forest (New Mexico), 102, 197–98
Gimiwan (Anishinaabeg native), 199–201, 244

Glacier Peak Wilderness Area (Washington), 15
globalization, 25, 229
global warming. *See* climate change and global warming
GMOs (genetically modified organisms), 48, 230–31, 235
grain agriculture, 234–37, 240
Grand Canyon National Park, 141
Grand Teton National Park, 17
grazing: in Arizona, 79; in Bandelier Wilderness, 85; beneficial effects of, 83, 107–8; in Chaco Culture National Historical Park, 81–83; controlled grazing, 46; in the Coon Creek watershed, 34; and the DBOC, 107–8; and ecological renewability, 80; and environmental damage, 52, 79, 83, 85, 208–12; in Gila Wilderness Area, 102–3; human dependency on, 249; as land management, 237; mob grazing, 34, 43–44; as natural process, 78; and the Wilderness Act, 85, 103; in wilderness areas, 21. *See also* overgrazing
Great Awakening, 233–36
Great Migration, 143, 173
Great Old Broads for Wilderness (environmental activist organization), 21
Greencorps Chicago, 146–54, 245
*Green Fire* (film), 197
greenhouse gases, 216, 219, 233–34
Green Revolution, 234–35
Grignon, Jeff, 67–74, 244, 262
Grim, John, 225n1
groves. *See* forests

habitat(s): critical habitats, 80; domestic habitation, 48; of the Drakes Estero system, 108; human habitats, 13, 250–51; in managed lands, 172; of the Rarámuri, 24; replenishment of, 92; urban habitat, 6, 141, 156–57
Handshake Agreement (1932), 22
Harrell, Stevan, 98n4
Harris, Nyeema, 192
Haugen, Joseph and Ernest, 33–36, 39
Hausdoerffer, Atalaya, 243–46, 248–49, 254
Hausdoerffer, John, 195–204, 243–54, 262
Hausdoerffer, Sol, 246
Hawken, Paul, 52
"Healing the Urban Wild" (Van Horn), 145–55
Hecht, Brooke, 205–13, 245, 262
Hedeen, Stan, 181–82
Hegewisch Marsh (Chicago), 153–54
Hendee, John C., 16
herons, 161–62
Higgins, Margot, 113–22, 245, 251, 262
High Line Canal (Colorado), 14–15, 19
High Sierra camps (Yosemite National Park), 105
Hills, Justin, 192
*H is for Hawk* (Macdonald), 19
Ho-Chunks (American Indian nation), 40
*Homo generativus* (generative man), 224–25, 225n1
*Homo modernus industrialii* (modern industrialized man), 29, 32
*Homo sapiens* (wise man; modern man), 17, 28–29, 196
Honor the Earth (Native-led environmental organization), 196
Howard, Amelia, 168–69, 172, 175
Howard, Michael, 166–76, 245, 263
human/nature continuum: in acequia communities, 91–92; awareness of, 218–19, 239; and bacteria, 231; as a biome, 231; and biomim-

icry, 50–51, 56–59; coexistence in, 22; colonialism vs., 97; degrees of wildness in, 4–5; in Ghana, 189–92; vs. human domination, 142; human impact on, 17–18; Menominee conceptions of, 68–71, 74; Rarámuri conceptions of, 25–26, 28; reapperception of, 220; between seeds and cultivators, 230. *See also* separation from nature

humans: as beneficial to nature, 20, 50–59, 67–74, 96, 100–112, 113–22, 145–55, 166–76, 179; in conflict with animals, 158–61; as conquerors/dominators, 215–17, 224–25, 233, 248, 250, 252; ecological and human health, 235–36; excluded from nature, 4, 13, 19, 96–97, 100–112, 148, 230; as guardians vs. gardeners of the natural world, 247, 249, 253; human will, 246–47; psychological breaks in, and ecosystems, 208; self-willed humans, 203, 217; the wild's relation to, 180–81. See also *Homo sapiens*; keystone species; separation from nature

"Hummingbird and the Redcap, The" (Peña), 89–99

hummingbirds, 6, 89–92, 98

husbandry, 6, 183–84

Ice Age, 67–68, 74, 221–23

Iceland, 205–13, 213n1

India, 7, 228–32, 252

Indian Heaven Wilderness Area (Washington), 22

Indian Peaks Wilderness (Colorado), 13

Indian Service (US), 27

individualism, 94–95, 250

"Inhabiting the Alaskan Wild" (Higgins), 113–22

injustice, 20–21, 168. *See also* justice

interconnectedness of humans and nature. *See* human/nature continuum

Interface, Inc., 51, 53–55, 58

*Interpreting Indicators of Rangeland Health* (US government publication), 81

"Into the Wildness" (Van Horn), 1–8

invasive species, 81, 86, 147, 149, 154

"Island Civilization" (Nash), 7, 243, 249–53. *See also* "Earth Island"

Jackson, Wes, 233–41, 249–50, 263

Jackson Hole, Wyoming, 17

Jazmin (Chicago ecological intern), 161–64

Jeffers, Robinson, 60–61, 265

Jim Crow era, 173

Johnson, Samuel (Dr. Johnson), 16

Jones, Bill, 53, 55, 59

Jordan, Henri, 146–47, 152–53

justice, 196–97, 201, 235–37, 247, 264, 267. *See also* injustice

Kahn, Peter, 17

Kaibab National Forest (Arizona), 79–81

Kansas State University, 239

Kansas Wesleyan University, 239

*Kenew* (Menominee eagle spirit), 67–69, 73

Kennecott industries, 15, 119

Kernza (grain variety), 236, 238

keystone species: and American Indians, 29; buffalo as, 202; humans as, 5, 7, 24, 28, 32, 94, 162, 181, 247

Kimmerer, Robin Wall, 67–74, 244, 263

kincentric ecology, 5

kincentric landscapes, 93

Kojève, Alexandre, 96

Kolyma (Siberia), 221–22

Komunyakaa, Yusef, 189, 194

LaDuke, Winona, 195–203
Lamont (Eden Place enthusiast), 170–71
"Land Ethic, The" (Leopold), 77, 92–93, 196–97
land health: assessment of, 82; awareness of, 217; and diversity, 81–83, 87; and economics, 87–88; and environmentalists, 75–84; and land use, 86–88; Leopold, Aldo on, 1, 78, 80–81, 84–85, 198; reconsidered, 79; and self-renewal, 78, 93, 200–201, 255; value-function controversy over, 83–86; and water, 76, 80–82, 85–87, 148; and wildlife, 34, 46–47, 87. *See also* rangeland health
land illiteracy, 80
Land Institute (Kansas), 239, 249, 263
land management, 19, 24–25, 87, 102, 118, 197
land ownership, 6, 38, 82, 95–96, 106, 138, 141, 238–39
land resting, 81, 83, 87
land use: forced changes in, 38; and greenhouse gases, 233; and human domination, 95; humans excluded from, 109–12, 113–22; and land health, 86–88; natural conditions of, 34, 36; soil degradation resulting from, 233–35; and the Wilderness Act, 102. *See also* land health; Leopold, Aldo; National Park Service
*Last Common, The* (Peña), 98n2
*Last Standing Woman* (LaDuke), 196
leave no trace (wilderness recreation concept), 106
Lecomte, Henri, 222
Lee, Jon, 40–41
LEED (Leadership in Energy and Environmental Design) certification, 182
Leonardo da Vinci, 240

Leopold, Aldo: as apostle of the Land, 139; and automobile traffic, 102; and biotic arrogance, 16; and colonialism, 201; on community, 1–2; and Coon Valley, 34, 40; eco-centric wilderness concepts of, 201; on ecological progress, 75; on farming, 183, 198; Flint's opposition to, 17; "The Forest," 197; *Game Management*, 87; holistic views of, 86; and *Homo generativus*, 225n1; "The Land Ethic," 77, 92–93, 196–97; on land health, 1, 78, 80–81, 84–85, 198; on land use, 102–3, 198–99; and Pyle, 14; and roadless areas, 21; *A Sand County Almanac*, 84, 196; on wilderness, 33, 84; "Wilderness as a Form of Land Use," 198–99. *See also* "*Akiing* Ethic, The"
Lincoln Park (Chicago), 162
"Listening to the Forest" (Grignon and Kimmerer), 67–74
Little, Charles, 86
*Log from the Sea of Cortez* (Steinbeck), 17–18
logging, 14–15, 20, 38, 94–95, 181, 190. *See also* forests; timber production
Long Term Ecological Research (LTER) sites, 161
Lopez, Barry, 184
"Losing Wildness for the Sake of Wilderness: The Removal of Drakes Bay Oyster Company" (Watt), 100–112
Louv, Richard, 180
Lunny family, 101, 109

Maasis (American Indian nation), 246
Macdonald, Helen, 19
Macfarlane, Robert, 1
Magle, Seth, 156–65, 245, 263

Maple Tree Nations (in Menominee guardian legend), 67–69
margins/marginality, 7, 17, 142–44. *See also* ecotones; "On the Wild Edge in Iceland"; "Wild Black Margins"
Maroon Bells-Snowmass Wilderness Area (Colorado), 12, 243–45
Marshall, Robert, 14, 16, 180
Marx, Karl, 93, 98n3
McDonough, William, 54
McKibben, Bill, 215–17
Meeker, John, 179
Meezan, Erin, 59
Meine, Curt, 33–42, 197, 248, 264
Menominee Forest, 6, 67–71, 73–74, 262
Menominees (American Indian nation), 68–71, 262
Mesoamericans, 95, 98n5
Mill Creek (Cincinnati), 181–82
mining, 20, 94–95, 103, 110, 116, 118–20
Mississippi River, 34, 36
mob grazing, 34, 43–44
monarch butterflies, 219, 230
monkeys. *See* colobus monkeys as gods
monocultures, 34, 45, 47, 70, 94, 216, 237
Moore, Curtis, 148, 152
mountaintop removal, 18
Mount Airy Forest (Cincinnati), 181, 184
Muir, John, 83, 86, 139, 178–79, 184

Nabhan, Gary, 19
Nader, Ralph, 196
Nash, Roderick Frazier, 14, 200–201, 243–54
National Academy of Sciences (US), 101
national parks: allure of, 28; automobile traffic in, 102; Banff National Park (Canada), 164; designations of, 120; functions of, 172; Grand Canyon National Park, 141; Grand Teton National Park, 17; humans excluded from, 113–21; hunting in, 21; increase in, 115; mining prohibited in, 119; as protected areas, 76, 87, 172, 184; residents in, 115–16; tourism in, 118–19; wilderness in, 103, 105; Wrangell-St. Elias National Park and Preserve, 20, 113–21, 251, 262; Yellowstone National Park, 25, 28, 180, 191, 246; Yosemite National Park, 105, 184
National Park Service (US): in Alaska, 115, 118–19; communicating with local populations, 120; differences of opinions within, 118–20; land uses allowed by, 20, 101–5, 115–16, 118–19; land uses predating, 107, 115; limitations on, 104; mining prohibited by, 119; and pristine wilderness, 107. *See also* Drakes Bay Oyster Company
National Riparian Team, 81
Native Americans, 19, 45, 95, 115–16. *See also* American Indians; First People
Native cultures and people: of Alaska, 115–16; being, as shared concept among, 95, 98n5; colonialism vs, 98; displacement of, 121; extermination of, 230; and forests, 70; going native, 92; in Iceland, 206; and wildlife conservation, 16
nature-based lifestyles, 29
Nature Boardwalk (Lincoln Park, Chicago), 162
Nature Conservancy, 20
nature deficit disorder, 180
Navajos (Native Indian nation), 31, 82–83
New Mexico, 75–79, 81–85, 197–98

niche-abiding life, 92–97
Nirvana (rock band), 18
nitrogen, 72, 207–12, 230, 238
Niwot Ridge study site, 13
North American Wildlife Conference (seventh), 1–2
"Notes on 'Up at the Basin'" (Rothman), 60–63
*Not Man Apart* (Brower, ed.), 60
"No Word" (Salmon), 24–32
NPS. *See* National Park Service (US)

Oakey, David, 51–55, 59. *See also* Interface, Inc.
oceans, 31, 38, 219, 221
"Ode to the Maggot" (Komunyakaa), 189
oil, 57–58, 218, 229, 234, 236, 238. *See also* fossil fuel development and consumption
Ojibwes (Native American nation), 37, 202
Oliver, Mary, 178, 183
oneness with nature. *See* human/nature continuum; separation from nature
"On the Wild Edge in Iceland" (Hecht), 205–13
operating conditions of the Earth, 53–54, 58
"Order versus Wildness" (Salatin), 43–49
Oregon Fish and Wildlife Commission, 20
Organ Pipe National Monument, 19
overgrazing, 76–77, 79, 82–83, 87
overpopulation. *See* population growth
Owen (Greencorps Chicago crew member), 151

Paleozoic Plateau. *See* Driftless Area
Palo Alto Ranch (New Mexico), 76
pandas, 191, 193–94

"Panther at the Jardin des Plantes, The" (Rilke), 214–23
partnership. *See* "Epilogue: Wild Partnership"
pasture cropping, 45–46
paternalism, 247–51
Peña, Devon G., 89–99, 252, 264
People's Climate March (New York City), 217, 224
permaculture, 47. *See also* biomimicry
pigs, 46–47, 49, 174, 181, 206
pilgrimage, 6, 141, 184
Pinchot, Gifford, 197
Pine Nation (in Menominee guardian legend), 67–69
*Place in Space, A* (Snyder), 145
Plato, 240
Point Reyes (California), 100–101, 104–6, 246
Pollan, Michael, 109, 183, 265
pollution, 31, 153, 171, 181–82
Polly Dyer Cascadia Broadband (environmental activist group), 21
Polyface Farm (Virginia), 5–6, 47, 265
population growth, 13, 25, 52, 121, 233, 250
Potts, Rick, 225n1
*Practice of the Wild, The* (Snyder), 177
practices of the wild. *See* "Toward an Urban Practice of the Wild"
pre-Columbian land management, 24–25
pristine wilderness. *See* wilderness
Pyle, Robert Michael, 12–23, 244, 264

Quinn, Daniel, 52
Quitobaquito Springs (Arizona), 20

racism, 140, 146, 168
ranches, 6, 76, 77–79, 87
rangeland health, 80–84. *See also* fires and fire; grazing; overgrazing

*Rangeland Health* (Gadzia), 81
Rarámuri (indigenous Mexican population), 3, 5, 24–27, 30–32, 265
Raven, Peter, 86
Ray, Shann, 16
reconciliation ecology, 160
*Recovering the Sacred* (LaDuke), 196
recovery: and grazing, 44, 78, 82; through nature, 145, 151; *Recovering the Sacred*, 196–97; White Earth Land Recovery Project, 196; of wildness, 5, 7
recycling, 38, 51–53, 56, 58, 152. *See also* waste
regeneration, 6, 70–74, 86, 92, 221–23, 244, 262
Reiger, H. C., 229
relative wild: in cities, 164, 181; as a continuum, 21; and the cultivated wild, 155; and degrees of wildness, 4–5; extinction of, 192; vs. the idealized wild, 109–10; inclusive attitudes toward, 18–19; and "Island Civilization", 252–53; and land health, 87; and Leopold, 199; Nash vs., 249, 253; in urban areas, 180–81; and the Wilderness Act, 101
renewable energy, 229, 231. *See also* solar energy/power
restoration, 84, 87, 182
rewilding, 104, 193
rice and rice harvesting, 7, 69, 196–203, 204n1, 237
*Rights of Nature, The* (Nash), 247
Rilke, Rainer Maria, 214–21
riparian areas, 38–39, 47, 81, 91–92, 96
rivers: in Alaska, 117; in Colorado, 14, 19; Culebra River, 90–91, 94, 96, 98n1; in the Driftless Area, 36, 39; in ecosystems, 31; Ganges River, 228–29; in "Island Civilization," 252–53; Kolyma River, 221; land uses prohibited in, 20; Los Pinos River, 130, 134; in the Menominee nation, 68; Mississippi River, 33, 36, 41; as sacred, 192, 228; in Uganda, 192
roadless areas, 16, 21, 102–3
Rockefeller family, 17
Rodin, Auguste, 214
Romantic movement, 240
Rothman, David J., 60–63, 248, 251, 264–65
Round Lake (Minnesota), 196–99
Roundup (pesticide), 230–31

Salatin, Joel, 43–49, 248, 265
Salmón, Enrique, 3, 24–32, 93, 244, 265
*Sand County Almanac, A* (Leopold), 84, 196
Sangre de Cristo Mountains (Colorado), 89
San Luis Valley (Colorado), 89–91, 94–95, 252
Savory, Allan, 81–82
Sayre, Nathan, 118
Seis, Colin, 45–46
self-ablazeness, 1–2, 7
self-renewal: of community, 1–2; disturbances of, 215–16; human promotion of, 247; and land health, 78, 93, 200–201, 255; in wilderness, 244
self-will, 2–3, 7, 150, 201, 218, 246–50, 252–53
self-willed/willing natural systems: and acequias, 92–98, 98n2; climate change threats to, 98n2; disappearance in, 109–10; human habitation in, 250–51; in "The Panther," 218; and self-renewal, 200–201; and wildness, 90, 93
separation from nature: and Agamben, 98n6; awareness of, 218–19, 240; and capitalism, 97; colonial projects enforcing, 97–98; grain agriculture

separation from nature (*continued*)
vs., 240; and the meaning of wild,
3, 28–29; resisted, 93–94; as a spiritual danger, 202–3. *See also* human/nature continuum
Sequoia-Kings Canyon (California), 105, 110n1
Serafin (Abeyta's great-grandfather), 125, 132
Shakti (self-organizing power), 7
sheep, 123–28, 130, 134, 206–7, 210–12, 237–38
Shiva, Vandana, 228–32, 252, 265–66
Sierra Club, 60, 105–7, 148, 181
Sierra Madre mountains (Mexico), 24, 198
Sierra Tarahumara (Mexico), 24–25
*Silene* (flowering plant species), 222–24
*Silphium* (sunflower relative), 236, 238
Silver City, New Mexico, 75–76
silviculture. *See* forests
slaves and slavery, 27, 40, 173, 219, 244, 247–50, 253
Smith, Frederick, 111n5
Smith, Huron, 71
Smith, Jeffrey, 48
Smith, Mistinguette, 137–44, 251, 266
Snaefellsjokull glacier (Iceland), 206
Snow Basin (Massachusetts), 60–63
Snyder, Gary: as contributor, 266; on cultural restoration, 145; and the human/nature continuum, 18; *A Place in Space*, 145; *The Practice of the Wild*, 177; salutation of, 5; "Wildfire News," 11, 18; on wildness and the wild, 2–4, 150, 198, 200–201
soil: abuse of, 34; acequia replenishment of, 92; bacteria in, 164, 178, 209; in the Bandelier Wilderness, 85; black people's relationships with, 143; carbon sequestration rates of, 238; conservation of, 35, 37, 40; contamination of, 168, 182;

of Coon Valley, Wisconsin, 34–35; degradation of, 233–35; of the Driftless Area, 37–40; of Eden Place, 173–76; erosion of, 35, 80–87, 91, 121, 140, 198, 234–35, 249; fertility of, 216; of forest floors, 71, 207–13; health of, 39, 45, 71–73, 75–76, 80–83, 207–13; microbiome management of, 238; nutrient cycling in, 78; and ordered farming, 45–46; plants' dependence on, 211; restoration of, by ranching, 87; self-renewing relationships with, 215–16; stability conditions for, 76; as unrenewable resource, 234; of White Earth Anishinaabeg Reservation, 198, 200; and wildness, 149, 238. *See also* "Listening to the Forest"; "Working Wilderness, The"
Soil Conservation Service, 34–35
solar energy/power, 116, 174–75, 197–98
Southern California Edison, 105
species extinction, 18–20, 22–23, 37, 216, 222, 230, 232, 248
Stark, Clem, 21
Stegner, Wallace, 84–85, 88
Steinbeck, John, 17–18
Steiner, Rudolph, 139
Steppenwolf (rock band), 3
"Story Isn't Over, The" (Warren), 214–27
Student Conservation Alliance (SCA), 152
Superstorm Sandy, 217
Supreme Court (US), 95, 174
sustainability: of agriculture, 107; through biomimicry, 57; Caldwell Preserve as model of, 181; corporate responses to, 51–54, 57–58; cultural sustainability, 71–74; of the DBOC, 108; of economic productivity, 35; of ecosystems, 28, 80; of food oper-

ations, 246; and garden ethics, 183; of grain agriculture, 237–38; human assumptions vs., 57; of land and water use, 20, 25, 35; of lifestyles and livelihoods, 6, 24; in the Menominee Forest, 69, 71; opponents of, 20; as a process, 183; and prosperity, 69; in urban areas, 174, 184; and wilderness, 179. *See also* unsustainability
Sutter, Peter, 17
Swimme, Brian, 225n1

taiga (Russia), 18
Tallmadge, John, 177–85, 245, 266
tar sand, 18
Taylor, Jack, 98n1
Taylor, Michael, 97
Taylor, Zach, 148, 151, 153–54
Taylor-Goodrich, Karen, 110n1
Thatcher Era, 19
Thoreau, Henry David: environmental mythology of, 179; on farming, 45, 183; and life pasturing freely, 15; portrayals of, 21; and slavery, 253; on walking in nature, 180; on Wildness, 2, 45, 61
*Thunder Tree, The* (Pyle), 15–16
timber production, 69, 71, 74, 103. *See also* logging
TNC (The Nature Conservancy), 20
Tohono O'odham (native Indian nation), 19
Toltec Unit (Colorado), 134
Toltec Wilderness (Colorado), 123
"Toward an Urban Practice of the Wild" (Tallmadge), 177–85
Tucker, Mary Evelyn, 225n1
Turner, Jay, 103

Uganda, 216, 219
UNESCO World Heritage sites, 81
United States, 84, 94–95, 115, 121, 142, 157, 219

United States Dairy Association (USDA), 174
United States government, 72–73, 79, 82. *See also* Congress (US)
University of Kansas, 238
Unknown, the (Scottish forest area), 23, 244
unpossessive presence, 92
unsustainability, 22, 57, 84–85
"Up at the Basin" (Rothman), 62–63
urban ecosystems, 6, 159–60, 162, 164–65, 168, 171, 176
urban sprawl, 75, 157, 166
urban wildlife. *See* wildlife (urban)

Vail, Mark, 113–15, 120, 251
value-function controversy, 83–86
Van Horn, Gavin, 1–8, 145–55, 266–67
Verne, Jules, 206
Vikings, 206–7, 212
Vogt, Kristiina, 208

Wahkiakum County (Washington), 20
Warren, Julianne Lutz, 214–27, 245, 267
Washington Fish and Wildlife Commission, 20
waste, 51, 54–57, 91, 168
"Wasted Wilderness" (Flint), 17
"waste equals food" concept, 54
water: for Chicago's wildlife, 161; cleaning and filtering of, 58; conservation of, 76; cycling of, 80; democracy of, 98n2; and dewatering, 19; in the Driftless Area, 37–40; among Earth's Operating Conditions, 53, 58; and erosion, 216; as farming's lifeblood, 47; and Ganges River hydrology, 228–29, 231–32; from glaciers, 119; in grain agriculture, 235; in Hegewisch Marsh, 154; of the High Line Canal, 14–15, 19; humans as beneficial to, 96, 101; incomplete control of, 91, 94, 96, 98n2;

water (*continued*)
  in industry, 58; and land health, 76, 80–82, 85–87; in the Menominee Nation, 68–69; and permaculture, 47; pollution of, 153, 182, 192–93, 234; protection of, 241; in rangelands, 76, 80; stability conditions for, 76; sustainable use of, 20; in the urban wild, 159–61, 169, 182; as wild area lifeblood, 47. *See also* acequia waterways; "*Akiing* Ethic, The"; Coon Valley; "Cultivating the Wild"(Shiva); Drakes Bay Oyster Company; "Inhabiting the Alaskan Wild"; oceans; rivers
watersheds, 34–35, 37–40, 76, 91–92
Watt, Laura Alice, 100–112, 244, 267
weeds: benefits of, 43–44, 46, 48; and grain agriculture, 240; invasive, 86, 149; and Roundup, 230; in urban areas, 143, 151, 171, 178, 199; weed trees, 73
Weston, Edward, 60
wetlands, 15, 91, 153, 167, 169–70, 182, 199
Whately, Booker T., 139, 144n1
"Whiskered God of Filth, The" (Dunn), 189–94
White, Courtney, 75–88, 248, 267
White Earth Anishinaabeg Reservation (Minnesota), 195–201
White Earth Land Recovery Project, 196
Whitman, Walt, 178
wild (adjective form): applied to American Indians, 29; and *autopoiesis*, 230; definitions and meanings of, 147–50, 173, 201, 203, 205; relatively wild nature, 180; the universe as wild, 61; wild agricultural processes, 237; wild black knowledge, 143; "Wild Black Margins," 137–44;

wild cities, 164; *Wild Earth*, 84; wild journeys, 5–7; wild practice of nature, 184; wild processes, 3, 7, 198, 202, 241; wild systems, 6–7, 45, 47, 52, 147, 237; wild uprisings in the Anthropocene, 217
wild (indeterminate form): absent from the Rarámuri language, 24, 26–27; in the colonized world, 230; definitions and meanings of, 2–3, 5, 12–13, 28, 31, 148–49, 200, 218–19, 244–46; reconceptualization of, 31–32
wild (the wild): in the Anthropocene, 215–17; the Big Wild, 16; and biomimicry, 50–51; businesses reconciling with, 56–59; colonial memory of, 138; complicated wild, 148–50; definitions and meanings of, 12–13, 177; and East Indian farming, 230; emergent wild, 150–55; extinction in, 18, 22–23; and Fernald, 183–84; human choices for, 254; human conquering/domination of, 39, 236, 247; humans excluded from, 7, 148, 173, 230, 249–50; idealized wild, 109; in Jeffers's poems, 61; killing the wild, 230–31; as managed lands, 172–73; Menominee concepts of, 70; operating conditions of, 54, 56, 58; in "The Panther at the Jardin des Plantes," 214; planetary wild, 6–7; practices of, 6, 177–85; relational wild, 250; in relation to humans, 180–81; sacred aspects of, 192; and socioeconomic class, 173; urban children encountering, 171; and Wilderness, 13–14; wisdom of, 5–6. *See also* "Building the Civilized Wild"; "Cultivating the Wild" (Shiva); "Cultivating the Wild on Chicago's South Side"; "Healing the

# INDEX

Urban Wild"; relative wild; "Wild Black Margins"; wildness
"Wild Black Margins," 137–44
wilderness: African Americans' experiences of, 137–43; ancestral wilderness, 195–204; awareness of, 84; and colonialism, 244; definitions and meanings of, 12–13, 84, 102–3, 200, 218, 244, 246; as a democratic ideal, 17; deterioration of, 84–85; endangered wilderness, 250; human conquering/domination of, 252; as human habitat, 13; human livelihood in, 67–74, 115–21, 244; and the human/nature continuum, 22; humans excluded from, 13, 244, 249–52; internal wilderness, 140; laws/legislation on, 20–21, 84, 101, 102–3, 106, 114–16; Leopold on, 33, 84, 197–99; living in, 121; marine wilderness, 100–112; and the Menominee Forest, 68, 69; mutualism in, 248; Native vs. Western views of, 201; necessity of, 180; opposition to, 16–17; potential wilderness, 6, 101, 104–7, 111; pristine wilderness, 37–38, 86, 101, 104, 106–7, 109–10, 179, 211, 245, 248–49; protection of, 16, 76, 86–87; regeneration in, 244; as sacred, 244; and self-will, 252; and time, 179; Toltec Wilderness, 6, 123; in urban areas, 161; vehicles in, 13, 17, 20, 22, 102, 105–7; as wholeness, 150; and wildness, 100–112, 179, 244, 246. *See also* wilderness areas
Wilderness (proper noun form), 13–14, 16–19, 22, 106, 244
Wilderness Act (1964): activities allowed in, 21; in Alaska, 115; commercial operations prohibited by, 100–112; creation of, 14, 21; and the debate over wilderness, 246; fiftieth anniversary of, 4; forward-looking nature of, 22; and local industries, 100–112; and mining, 103, 110; passage of, 84; and the relative wild, 101; wilderness defined in, 84, 102; and the Wilderness Society, 102; Zahniser as primary author of, 247
*Wilderness and the American Mind* (Nash), 14, 244, 246
wilderness areas: commercial operations prohibited in, 107; designated wilderness areas, 45, 102–5, 104–5, 121; Federal Wilderness Areas, 13, 17, 22, 84; functions of, 172; grazing in, 21; Hausdoerffer on, 203; humans excluded from, 97, 100–112, 248; Leopold on, 84; logging in, 102–3; as managed lands, 172; Point Reyes Wilderness Area, 100–112, 246; as unreal, 29; unsustainability in, 84–85; in Washington, 12. *See also* Wrangell–St. Elias National Park and Preserve
"Wilderness as a Form of Land Use" (Leopold), 198–99
"Wilderness in Four Parts, or Why We Cannot Mention My Great-Grandfather's Name" (Abeyta), 123–34
Wilderness Society, 102–3, 107
"Wildfire News" (Snyder), 11
wildfires. *See* fires and fire
Wildlands Project (environmental activist group), 106
wildlife: balance in, 48; biology of, 35; conservation of, 230; depletion of, 38; and land health, 34, 46–47, 87; prospects for, 85–86; protection of, 80; returning to rural areas, 238; road crossings for, 164; as sacred, 189–94

wildlife (urban). See "Building the Civilized World"; "Cultivating the Wild on Chicago's South Side"; "Healing the Urban Wild"; "Toward an Urban Practice of the Wild"; "Wild Black Margins"

wildlife refuges: acequia farms as, 90; biodiversity in, 172; bordering ranches, 76, 79; Buenos Aires National Wildlife Refuge, 76; in Chicago, Illinois, 168–69; ecological health of, 76, 79–80, 87; functions of, 172; as managed lands, 172; overgrazing in, 87; as protected areas, 76; Stegner on, 85; Wilderness Act provisions for, 103

wildness: acequias as extensions of, 92; and agriculture, 47, 49, 235–36; allure of, 28, 180; ancestral wilderness, 201; black people associated with, 138; in Chinese art, 1; circling wildness, 2–4; definitions and meanings of, 2–3, 201, 244–47, 250; degrees of, 4, 33, 145–46; and disease, 49; in DNA, 48; as Earth's operating condition, 53; Fernald's recovery of, 183–84; and fossil fuels, 229; and genius of place, 51; healing processes of, 151–53, 172, 176; human conquering/domination of, 48–49, 57, 240; and the human/nature continuum, 14–15; human restoration of, 179; humans excluded from, 147–48, 249–52; of Iceland, 212; internal balance in, 47; Leopold on, 201; order vs., 43–49; patterns innate to, 49; pervasive nature of, 4, 15, 21–22; as practice, 3–4; predator-prey balance in, 48; processes of, 3; questions raised by, 5–6; reconceptualization of, 31; recovery of, 5, 7; relationality of, 90, 96; as restoration of people and place, 150, 176; and sacred groves, 191, 193; of seeds, 230; and the self, 90; as self-organizing and renewing system, 177, 179, 244; and self-will, 93, 150, 250; serenity in, 150; Snyder on, 177; in soil, 238; and time, 18, 179; as urban violence, 148–49; urban wildness, 181, 183; and wilderness, 100–112, 179, 244, 246; as Wildness, 61–62; "The Working Wilderness," 75–88; working wildness, 104–6; of the world, 2

*Wild Places, The* (Macfarlane), 1
Willapa Hills (Colorado), 15
Williams, Terry Tempest, 2
Wilson, E. O., 29, 178, 180, 251
wisdom of the wild, 6
working wild, 6
"Working Wilderness, The" (White), 75–88
Wornum, Michael, 106
Worster, Don, 240–41
Wrangell-St. Elias National Park and Preserve (Alaska), 20, 113–21, 251, 262
Wright, Angus, 235

Yakamas (Native Indian nation), 22
Yake, Bill, 21
Yale School of Forestry and Environmental Studies, 197, 209–10, 262
Yellowstone National Park, 25, 28, 180, 191, 246
Yukaghirs (Native American nation), 222

Zahniser, Howard, 21, 103, 247, 253
Zwinger, Ann, 183